ORGANIC CHEMISTRY I

By the Staff of
Research & Education Association

Research & Education Association
Visit our website at
www.rea.com

Research & Education Association
61 Ethel Road West
Piscataway, New Jersey 08854
E-mail: info@rea.com

SUPER REVIEW®
OF ORGANIC CHEMISTRY I

Published 2011

Printed in the United States of America

Library of Congress Control Number 00-131316

ISBN-13: 978-0-87891-192-9
ISBN-10: 0-87891-192-8

J10

(more on next page)

(continued from previous page)

"Your books have saved my GPA, and quite possibly my sanity. My course grade is now an 'A', and I couldn't be happier."

Student, Winchester, IN

"These books are the best review books on the market. They are fantastic!"

Student, New Orleans, LA

"Your book was responsible for my success on the exam. . . I will look for REA the next time I need help."

Student, Chesterfield, MO

"I think it is the greatest study guide I have ever used!"

Student, Anchorage, AK

"I encourage others to buy REA because of their superiority. Please continue to produce the best quality books on the market."

Student, San Jose, CA

"Just a short note to say thanks for the great support your book gave me in helping me pass the test . . . I'm on my way to a B.S. degree because of you !"

Student, Orlando, FL

WHAT THIS Super Review WILL DO FOR YOU

This Super Review provides all that you need to know to do your homework effectively and succeed on exams and quizzes.

The book focuses on the core aspects of the subject, and helps you to grasp the important elements quickly and easily.

Outstanding Super Review features:

- Topics are covered in logical sequence
- Topics are reviewed in a concise and comprehensive mann
- The material is presented in student-friendly language that makes it easy to follow and understand
- Individual topics can be easily located
- Provides excellent preparation for midterms, finals and in-between quizzes
- In every chapter, reviews of individual topics are accompanied by Questions Q and Answers A that show how to work out specific problems
- At the end of most chapters, quizzes with answers are included to enable you to practice and test yourself to pinpoint your strengths and weaknesses
- Written by professionals and test experts who function as your very own tutors

Larry B. Kling
Chief Editor

CONTENTS

12 CARBOXYLIC ACIDS

13 CARBOXYLIC ACID DERIVATIVES

CHAPTER 1

Structure and Properties

1.1 Atomic and Molecular Orbitals

Atomic orbitals are arrangements of electrons around the nucleus of an atom. An electron occupies an orbital according to its energy content. In order of increasing energy, the orbitals are specified by the letters s, p, d, and f, within a given shell. The shells are also arranged in order of increasing energy and are assigned the letters K, L, M, etc.

In Figure 1.1, the shapes of some of these orbitals are shown.

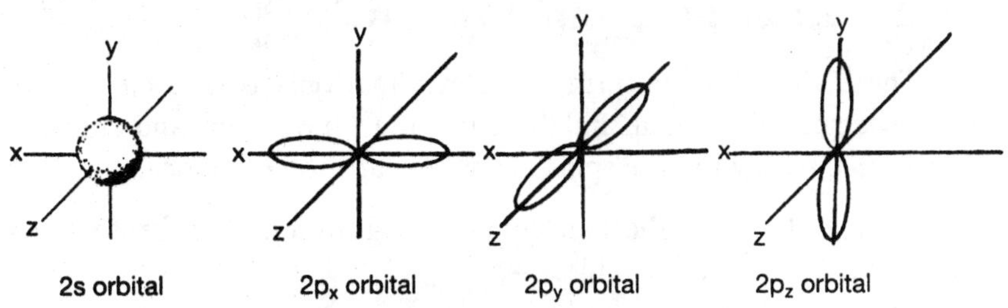

Figure 1.1 Atomic Orbitals (s and p)

The overlapping of atomic orbitals leads to the formation of molecular orbitals, and thus molecular bonding (covalent).

The sigma (σ) bond, with its characteristic shape, is formed from the overlapping of two s-orbitals, two p-orbitals, or an s and a p-orbital.

Two molecular orbitals, one bonding and one antibonding, are formed when two atomic orbitals are joined. The bonding orbital is of lower energy and is more stable than the component atomic orbitals. The antibonding orbital is of higher energy and is less stable than the component atomic orbitals. This is shown in Figure 1.2.

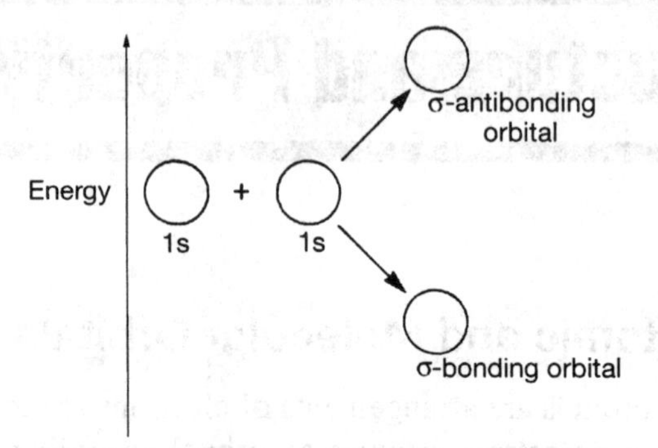

Figure 1.2 Formation of Two Molecular Orbitals

Electrons in antibonding orbitals lead to repulsive forces, which are almost as strong as the attractive forces in bonding orbitals.

1.2 Electron Configuration

The Pauli exclusion principle states that only two electrons can occupy an atomic orbital, and these two must have equal and opposite spins. Electrons with like spins cannot occupy the same orbital.

Table 1.1 shows the electronic configurations for the first ten elements of the periodic table.

Following the examples shown in the table, the electronic configuration of Argon is expressed as

$$1s^2 2s^2 2p^2{}_x 2p^2{}_y 2p^2{}_z 3s^2 3p^2{}_x 3p^2{}_y 3p^2{}_z$$

or equivalently

$$1s^2 2s^2 2p^6 3s^2 3p^6.$$

Table 1.1 Electronic Configurations

Symbol	Atomic Number		Electronic Configuration 1s	2s	$2p_x$	$2p_y$	$2p_z$
H	1	1s	↑				
He	2	$1s^2$	↑↓				
Li	3	$1s^2 2s$	↑↓	↑			
Be	4	$1s^2 2s$	↑↓	↑↓			
B	5	$1s^2 2s^2 2p_x$	↑↓	↑↓	↑		
C	6	$1s^2 2s^2 2p_x 2p_y$	↑↓	↑↓	↑	↑	
N	7	$1s^2 2s^2 2p_x 2p_y 2p_z$	↑↓	↑↓	↑	↑	↑
O	8	$1s^2 2s^2 2p^2_x 2p_y 2p_z$	↑↓	↑↓	↑↓	↑	↑
F	9	$1s^2 2s^2 2p^2_x 2p^2_y 2p_z$	↑↓	↑↓	↑↓	↑↓	↑
Ne	10	$1s^2 2s^2 2p^2_x 2p^2_y 2p^2_z$	↑↓	↑↓	↑↓	↑↓	↑↓

Problem Solving Example:

Would you expect the hypothetical species He_2^+ to be more stable than the hypothetical species He_2? than He?

A This problem requires a consideration of bonding and antibonding orbitals. An atomic orbital may be thought of as a particular volume element at a given distance and direction from the nucleus where there is a probability of finding an electron. According to the Pauli exclusion principle, no more than two electrons can occupy an orbital. Two overlapping orbitals can be considered to combine to give one low-energy bonding orbital and one high-energy antibonding orbital.

Consider the electron configuration of a helium atom (He): $1s^2$. We can readily see that the outermost shell of helium, the s orbital, is filled with two electrons, which is the maximal and stable capacity. Thus, the helium atom is very stable. In fact, it is inert to most chemical reactions and is for this reason known as one of the inert gases.

To form the hypothetical species He_2, each helium atom contributes two ls electrons for a total of four electrons. Since the molecular bonding orbital can accommodate only two electrons, the remaining two must go into the antibonding orbital. Hence, any stabilization gained by filling the bonding orbital is offset by the required filling of the antibonding orbital. Hence, there exists no net decrease in energy when He_2 is produced, so that He_2 is less stable than He.

The only difference between the hypothetical species He_2 and He_2^+ is that the latter has only one electron (not two) in the antibonding orbital. It possesses two electrons in the bonding orbital as does He_2, but because of the one less electron in destabilizing antibonding orbital in He_2^+, it should be more stable than He_2. (It is, of course, less stable than He, which has no electrons in the antibonding orbital.)

1.3 Hybrid Orbitals

The sp hybrid orbitals arise from the mixing of one s orbital and one p orbital. These orbitals are equivalent and much more strongly directed than either the s or p orbital. The sp hybrid orbitals point in exactly opposite directions, which permits them to get as far away from each other as possible.

The sp^2 hybrid orbitals arise from the mixing of one s orbital and two p orbitals. These orbitals lie in a plane that includes the atomic nucleus. They are directed to the corners of an equilateral triangle with an angle of 120° between any of two orbitals.

The sp^3 hybrid orbitals arise from the mixing of one s orbital and three p orbitals. These orbitals are directed to the corners of a regular tetrahedron. The angle between any two orbitals is the tetrahedral angle 109.5°.

Problem Solving Example:

Assign all the electrons in acetaldehyde ($CH_3 - CH = O$) to their atomic or molecular orbitals.

A The formation of covalent bonds by the sharing of electrons results from the overlapping and interaction of partially filled atomic orbitals. The molecular orbitals (bonds) so formed are represented adequately by the summation of the geometrical properties of the individual atomic orbitals.

Since bond formation is produced by the interaction of orbitals, which can be highly directional, the most effective bonds will be formed when the relative spatial positions of the atoms are such as to produce the best possible overlap of orbitals.

To be able to assign all the electrons in $CH_3 - CH = O$ to their orbitals, first set up the electron dot formula:

```
     H    H
     ••   ••
  H:C¹: C²::Ö:
     ••
     H
```

When an atom forms a compound in which it is attached to four neighboring atoms by single bonds, sp^3 hybrid orbitals are used to form these bonds. Thus, carbon 1 is sp^3 hybridized. All hydrogen atoms have s orbitals; therefore, the three carbon-hydrogen bonds on the first carbon atom are $sp^3 - s$ bonds.

The carbon-carbon bond between the first and the second carbon are $sp^3 - sp^2$ bonds. sp^2 orbitals are utilized only in the formation of molecules containing double bonds. This gives a trigonal arrangement (planar) in space as opposed to the tetrahedral arrangement of the sp^3 hybrid orbital. Thus, the carbon-hydrogen bond on the second carbon atom is an $sp^2 - s$ bond, while the double bond on the oxygen atom is composed of two different bonds. One bond is the $sp^2 - p$ bond and the other (weaker) bond is a $p - p$ bond.

1.4 Chemical Bonding

An ionic bond is the electrostatic attraction between oppositely charged particles, which results from the transfer of electrons.

An ion-dipole bond is formed if one of the ions in an ionic bond is replaced by a highly polar molecule, such as water. This bond results from the attraction of the ion to the oppositely charged end of the polar molecule.

A dipole-dipole bond is formed if the ion in an ion-dipole bond is replaced with another polar molecule. This bond results from the attraction of the oppositely charged ends of the two polar molecules.

The ionic bonds form stronger bonds than ion-dipole bonds, which in turn are stronger than dipole-dipole bonds.

The formation of covalent bonds by the sharing of electrons results from the overlapping and interaction of partially filled atomic orbitals.

The degree of the overlapping of the atomic orbitals to form a bonding molecular orbital determines the strength of the covalent bonds.

Bond length is the distance between bonded nuclei. At this distance the repulsion that occurs between the similarly charged nuclei balances the "packing" effect of bonding.

Sigma (σ) bonds occur from the formation of σ-orbitals, which result from the end-to-end overlap of s and p atomic orbitals. It is also possible for the p orbitals to overlap side-to-side, in which case they form Pi (π) bonds. These bonds consist of π orbitals above and below the plane of the molecule. These are always present in double and triple bonds.

Problem Solving Examples:

Write a structural formula using a line for each of the substances represented below.

(a) $CH_3CH(CH_3)_2$ (b) CH_3CCCH_3

(c) CH_2ClOCH_2CHO (d) $(CH_2)_4$

(e) $CH_3CONHCH_3$ (f) $CH_2CH_2OCH_2CH_2O$

(a) In 2-methylpropane, $CH_3CH(CH_3)_2$, we have three hydrogens attached to the first and third carbon atoms. The methyl group consists of a carbon atom bonded with three hydrogen atoms.

We also have a hydrogen attached to the carbon in the second position. The second carbon is also bonded with three methyl groups.

```
    H   H   H
    |   |   |
H - C - C - C - H
    |   |   |
    H   |   H
      H-C-H
        |
        H
```

(b) CH_3CCCH_3: The first carbon has three hydrogen atoms and is bonded to the second carbon. The second carbon is only joined to the first carbon and the third carbon atom.

In order to complete an octet around the second carbon, it is absolutely necessary for the linkage with the third carbon atom to be triple bonded. The third carbon is attached to a methyl group.

```
    H               H
    |               |
H - C - C ≡ C - C - H        2-butyne
    |               |
    H               H
```

(c) CH_2ClOCH_2CHO: The first carbon is part of a methyl group and chlorine atom. It is attached to an oxygen atom. The last carbon is attached to a methylene group and an oxygen atom. Since the oxygen is not linked to any other atom and in order to be neutral, it must be doubly bonded to the last carbon.

```
          H          H        O
          |          |       //
Cl  -  C  -  O  -  C  -  C
          |          |       \
          H          H        H
```

(d) $(CH_2)_4$: Each carbon atom is attached to two other carbon atoms (in a ring) and two hydrogen atoms to form cyclobutane.

```
        H     H
        |     |
  H  -  C  -  C  -  H
        |     |
  H  -  C  -  C  -  H
        |     |
        H     H
```

(e) $CH_3CONHCH_3$: A methyl group is attached to the middle carbon. The latter has an oxygen atom, but since it needs to bond twice more, it should be bonded doubly to either nitrogen or oxygen. But since nitrogen (needing three bonds) is linked to hydrogen and a methyl group, it can only form a single bond with the middle carbon.

```
        H     O           H
        |     ||          |
  H  -  C  -  C  -  N  -  C  -  H
        |           |     |
        H           H     H
```

(f) $CH_2-CH_2OCH_2CH_2O$ (ring closure bracket joining first CH_2 and last O): In this compound the two oxygens are bonded with two $-CH_2-$ groups each, and the carbon atoms are linked to an oxygen, another $-CH_2-$ and two hydrogen atoms to form dioxane, a cyclic ether.

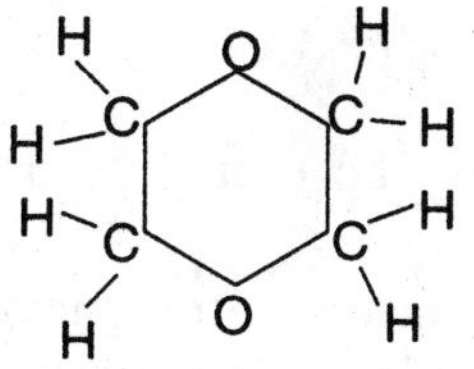

Q What is wrong with the following structures?

```
(a)     H              (b)      H H               (c)      H
        |                       | |                        |
      H-C-O                  H-C≡C-O-H                   H=C-O=H
        |                       | |                        |
        H                       H H                        H
```

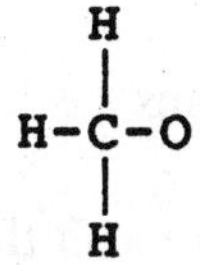

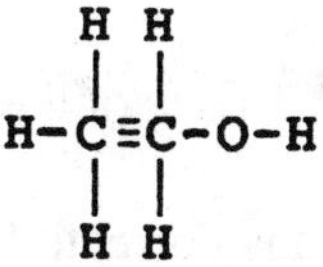

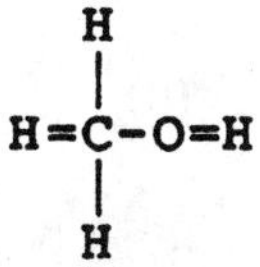

A Carbon, hydrogen, and oxygen each forms a specific number of bonds when they combine with other elements. This number is called their valence and is determined by their position in the periodic table. Carbon is in group IV and thus forms four bonds when it combines with other elements. Hydrogen is in group I and forms one bond. Oxygen is in group VI. It does not form six bonds, though. The actual number of bonds formed by any element is dependent upon the number of electrons in the outer shell of the atom. In nature, there is a rule that is followed that states that every element in combining with another element seeks to attain eight electrons in its outer shell. This is called the octet rule. Thus, hydrogen in group I will seek to combine with an element in group VII. This is because hydrogen has one electron to contribute and an element in group VII will have seven. Getting back to oxygen; it is in group VI and will, therefore, have six electrons to share. This means that it needs two more electrons to make up its octet. Oxygen needs to form two bonds to complete its octet.

In each one of the structures above, the error is in the number of bonds drawn to the various elements.

(a) H–C(H)(H)–O Taking the elements one by one, it is seen that each hydrogen atom forms one bond and is thus correct; the carbon atom forms four bonds, which is also correct, but the oxygen has only formed one bond. This indicates that there are only seven electrons in its outer shell and is not correct.

(b) H–C(H)(H)≡C(H)–O–H Each of the hydrogen atoms has formed one bond and is thus correct. The carbon atom on the left has formed three single bonds with the hydrogen atoms and a triple bond with the other carbon for a total of six bonds, which is incorrect. The carbon on the right also participates in the triple bond and in single bonds with the oxygen and hydrogen atoms. This is a total of five bonds and is also incorrect.

The oxygen atom has formed two bonds, which is correct.

(c) H=C(H)(H)–O=H Here, two hydrogen atoms are drawn with one bond and two are drawn with two bonds. The latter two are incorrect. There is a total of five bonds drawn to the carbon, which is incorrect. Carbon can form only four bonds. There are three bonds drawn to the oxygen atom. This is also incorrect because oxygen can form only two bonds.

1.5 Bond Dissociation Energy

The amount of energy liberated when a bond is formed is called the bond dissociation energy. This is also the energy needed to break the bond. The more energy that is lost when a bond is formed, the more stable the bond will be and the more energy will have to be used in breaking, or dissociating, it.

The breakage of a covalent bond resulting in an equal distribution of electrons to each of the fragments is known as homolysis.

$$A{:}B \rightarrow A\bullet + B\bullet$$

An unequal distribution of electrons is known as heterolysis.

$$A{:}B \rightarrow A{:} + B \text{ or } A + B{:}$$

Simple heterolysis of a neutral molecule yields a positive ion and a negative ion. In the gas phase, bond dissociation generally takes place by homolysis. In an ionizing solvent, heterolysis occurs more frequently.

1.6 Structure and Physical Properties

Polarity of Bonds

Polarity results from an unequal distribution of the electron cloud about two nuclei, such that the negative pole exists where the electron cloud is denser about one nucleus, and the positive pole exists where the electron cloud is sparser about the other nucleus.

A covalent bond will be polar if it joins atoms that differ in their tendency to attract electrons, that is, atoms that differ in electronegativity. The greater the difference in electronegativity, the more polar the bond will be. The following relates the electronegativity of some common elements: F > O > Cl, N > Br > C > H.

Examples of polar bonds are:

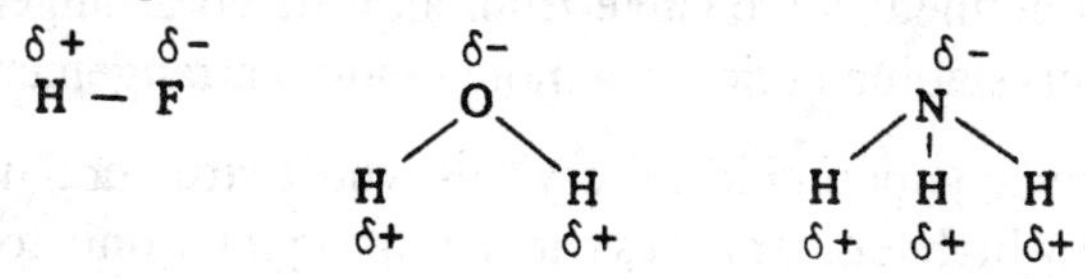

The symbols $\delta+$ and $\delta-$ indicate partial charges and are used to indicate polarity.

Polarity of Molecules

When the center of the negative charge of a molecule does not agree with the center of the positive charge, the molecule is polar.

A dipole is defined as a molecule with two equal and opposite charges separated in space. The dipole possesses a dipole moment, μ, which is equal to the magnitude of the charge, e, multiplied by the distance, d, between the centers of the charge:

$$\mu = e \times d$$

μ	e	d
in DEBYE units, D	in e.s.u.	in cm

The values of the dipole moments indicate the relative polarities of different molecules. The following are dipole moments for some molecules:

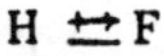

H ⟷ F

$\mu = 1.75D$

(H–C(H)(H)–H with bond dipoles)

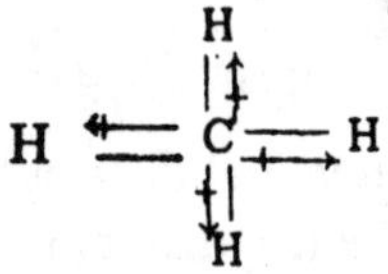

The dipoles cancel.

$\mu = 0D$

The symbol ⟶ represents a dipole, where the arrow points from positive to negative. Molecules, such as that of methane, with zero dipole moments, are nonpolar.

Melting Point

Melting occurs at the temperature where the kinetic energy of the particles is great enough to overcome the forces binding the particles. Melting is defined as a change from an ordered arrangement of particles in a crystalline lattice to a more random arrangement in a liquid.

An ionic compound forms crystals whose structural units are ions. Only at very high temperatures can the strong interionic forces be overcome.

In a nonionic compound, the atoms are held together by covalent bonds and form crystals in which the structural units are molecules.

Nonionic compounds melt at lower temperatures than ionic compounds due to weak intermolecular forces.

Intermolecular Forces

Dipole-dipole interaction is the attraction of the positive end of one polar molecule for the negative end of another polar molecule. Polar molecules are held more strongly together than are nonpolar molecules of comparable weight.

An example of a strong dipole-dipole interaction is hydrogen bonding, in which a hydrogen atom serves as a bridge between two electronegative atoms, holding one by a covalent bond and the other by purely electrostatic forces. The hydrogen bond is indicated by the broken line as follows: H – F --- H – F.

For hydrogen bonding to be important, the hydrogen atom must be attached to a highly electronegative atom, such as fluorine, oxygen, or nitrogen.

The forces between the molecules of a nonpolar compound are called van der Waals forces. These forces result from the formation of small momentary dipoles that are created from the instantaneous asymmetrical distribution of electrons and the attraction of these to similar momentary dipoles. These forces have a very short range in that they act only between portions of different molecules in close touch. Every atom has an effective "size" called its van der Waals radius.

Boiling Point

Boiling occurs at the temperature where the kinetic energy of the particles is great enough to overcome the cohesive forces that hold them in the liquid state.

The ion is the basic unit of an ionic compound in the liquid state. In order for a pair of oppositely charged ions to break away from the liquid, a great amount of energy is required. For this reason boiling occurs at very high temperatures.

The molecule is the basic unit of a nonionic compound in the liquid state. Boiling for nonionic compounds occurs at much lower temperatures than for ionic compounds because of the relatively weak intermolecular forces that must be overcome.

Associated liquids are liquids with molecules held together by hydrogen bonds. Because of the strength of these bonds, these liquids also boil at high temperatures.

Since molecular size is proportional to the strength of the van der Waals forces, polarity, hydrogen bonding, and thus boiling points, an increase in the size of the molecule corresponds to an increase in these properties.

Solubility

In order for an ionic compound to dissolve, the solvent must be able to form ion-dipole bonds and have a high dielectric constant. Only water or other highly polar solvents are able to dissolve ionic compounds appreciably.

Solubility of nonionic compounds is determined by their polarity. Nonpolar or weakly polar compounds dissolve in nonpolar or weakly polar solvents. Highly polar compounds dissolve in highly polar solvents. Nonionic compounds thus follow the rule, "like dissolves like."

Water is a poor solvent for most organic compounds. Solvents such as water and methanol are such examples. These are acidic solvents containing hydrogen atoms attached to oxygen or nitrogen atoms.

Problem Solving Examples:

What are the characteristics that ionic, ion-dipole, and dipole-dipole bonds have in common? How do they differ?

A bond between two species is defined in terms of the amount of energy liberated because of adduct formation. The strength of the bond increases as the amount of heat generated is increased at the expense of the total energy of the system.

An ionic bond is produced when an electropositive and an electronegative element react, by transferring electrons. For example, when

Na• and Na• and $:\dot{\underset{\cdot\cdot}{Cl}}:$ react, the electropositive sodium transfers one electron to chlorine; Na^+ and $:\ddot{\underset{\cdot\cdot}{Cl}}:^-$ are produced. This ionic bond is nondirectional.

An ion-dipole bond is generated when one of the ions in an ionic bond is replaced by a highly polar molecule such as water:

$$\overset{\delta+}{H}\text{—}\overset{\delta-}{O}\text{—}\overset{\delta+}{H} \equiv (+\;\;-)$$

When an ion such as Na^+ reacts with H_2O, the positive ion aligns with respect to the polar molecule as shown:

$$Na^+ \quad (-\;\;+)$$

Similarly, if another polar molecule replaces the ion in the ion-dipole bond, the formation of a dipole-dipole bond occurs. If we consider two polar molecules (e.g., two H_2O molecules), the alignment of the molecules is such:

$$\overset{\delta+}{H}\text{—}\overset{\delta-}{O}(\text{—}\overset{\delta+}{H})\cdots\overset{\delta+}{H}\text{—}\overset{\delta-}{O}\text{—}\overset{\delta+}{H}$$

The other hydrogen atoms also attract oxygen atoms in other water molecules.

Ionic bonds form stronger bonds than ion-dipole bonds, and the latter are stronger than dipole-dipole bonds.

Why does hydrogen fluoride undergo intermolecular self-association while the other hydrogen halides do not?

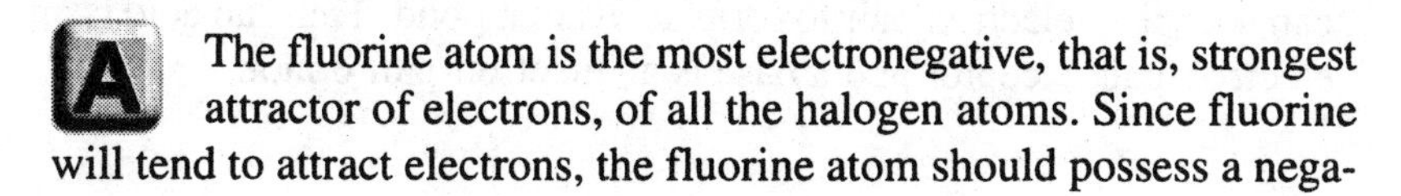

The fluorine atom is the most electronegative, that is, strongest attractor of electrons, of all the halogen atoms. Since fluorine will tend to attract electrons, the fluorine atom should possess a negative charge and the hydrogen atom a positive charge in HF. That is, HF is a polar molecule with one end (F) electron rich and one end (H) electron deficient. In a solution of HF, the opposite charges can attract each other, that is, one has intermolecular self-association. When opposite and equal charges are separated by some distance in a molecule (such as HF), a dipole is formed. Hence, the intermolecular self-association can be considered dipole-dipole interaction. This particular kind of interaction is also referred to as hydrogen bonding. It may be pictured as shown:

```
H-F ••• H-F
      ↑
      hydrogen bonding.
```

The tremendous electronegativity of fluorine is what allows the formation of hydrogen bonds in an HF solution. Since the other halogens possess less of a tendency to attract electrons, they will be less able to form hydrogen bonds required for intermolecular self-association.

1.7 Acids and Bases

The Lowry-Brönsted definition states that an acid is a substance that donates a proton and a base is a substance that accepts a proton.

The strength of an acid depends upon its tendency to donate a proton, and the strength of a base depends upon its tendency to accept a proton. The relative strengths of some common acids and bases:

$$\text{Acid strength} \quad \begin{matrix} H_2SO_4 \\ HCl \end{matrix} > H_3O^+ > NH_4^+ > H_2O$$

$$\text{Base strength} \quad \begin{matrix} HSO_4^- \\ Cl^- \end{matrix} < H_2O < NH_3 < OH^-$$

The Lewis definition states that a base is a substance that can donate an electron pair to form a covalent bond, and an acid is a substance that can accept an electron pair to form a covalent bond. Thus, an acid is an electron-pair acceptor and a base is an electron-pair donor.

The ability to accommodate the electron pair depends upon several factors, including:

a) the atom's electronegativity and

b) its size.

Within a given row of the periodic table, acidity increases as electronegativity increases:

Acidity $H - CH_3 < H - NH_2 < H - OH < H - F$

$H - SH < H - Cl$

Within a given family, acidity increases as the size increases:

Acidity $H - F < H - Cl < H - Br < H - I$

$H - OH < H - SH < H - SeH$

Problem Solving Examples:

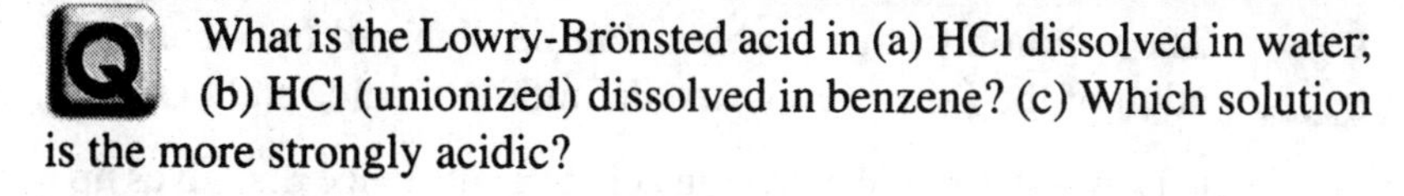

Q What is the Lowry-Brönsted acid in (a) HCl dissolved in water; (b) HCl (unionized) dissolved in benzene? (c) Which solution is the more strongly acidic?

A A Lowry-Brönsted acid may be defined as a substance that gives up a proton (hydrogen nucleus). A base, according to the Lowry-Brönsted idea, would then be a substance that accepts the proton.

(a) When HCl is dissolved in water, the water acts as a base due to the electron pairs available for sharing around the oxygen. Hence, the HCl can donate a proton (H^+) and protonate the oxygen atom in H_2O as shown:

$$HCl + H_2O \rightleftarrows H_3O^+ + Cl^-$$

The hydronium ion (H_3O^+) generated on the right is then the Lowry-Brönsted acid in solution, for it can donate the proton in the reverse reaction and hence acts as an acid.

(b) When unionized HCl is added to benzene, a hydrocarbon (that is, a compound containing only hydrogen and carbon atoms), the Lowry-Brönsted acid can only be HCl. No reaction occurs between the benzene and HCl to generate any other acid.

(c) The Lowry-Brönsted acid H_3O^+ is weaker than HCl. The strength of an acid depends upon its tendency to give up a proton. HCl gives up its proton very readily, whereas H_3O^+ has a greater tendency to hold onto its proton. The strengths of some common acids are indicated below.

$$\text{Acid strength } \begin{matrix} H_2SO_4 \\ HCl \end{matrix} > H_3O^+ > NH_4^+ > H_2O$$

Since HCl is the stronger acid, the HCl in benzene is the more acidic solution.

Q Account for the fact that nearly every organic compound containing oxygen dissolves in cold concentrated sulfuric acid to yield a solution from which the compound can be recovered by dilution with water.

The solution to this problem can be obtained by a consideration of the Lowry-Brönsted definition of acids and bases.

In the Lowry-Brönsted sense, an acid is a substance that gives up a proton, and a base is a substance that accepts a proton. When the cold concentrated sulfuric acid is added to the organic compound containing oxygen, it gives up a proton (hydrogen nucleus) and protonates the oxygen atom, which possesses electron pairs available for sharing. The reaction may be illustrated as shown:

$$-\overset{|}{\underset{\cdot\cdot}{O}}: + H_2SO_4 \rightleftarrows -\overset{|}{\underset{\cdot\cdot}{O}}{}^+:H + HSO_4^-$$

The oxygen atom of the organic compound is acting as the base because it accepts the proton from the sulfuric acid. It is thus protonation, which is reversible, that accounts for the organic compound dissolving in the cold concentrated sulfuric acid.

The protonation generates (as shown previously) the following:

$$\underset{\textbf{A}}{-\overset{|}{\underset{\cdot\cdot}{O}}{}^+:H} \quad \text{and} \quad \underset{\textbf{B}}{HSO_4^-}$$

"A" is the new acid formed because it can donate the proton; "A" is a weak acid (it will tend not to give up the proton readily). "B" is the new base generated, and it is also weak (it will tend not to accept a proton donated). If water is added to this solution, the organic compound will be regenerated, because water is a stronger base than HSO_4^- so that it will accept the proton from "A." The reaction can be written as:

$$-\ddot{O}{:}H^+ + H_2O \rightleftarrows -\ddot{O}{:} + H_3O^+$$

organic compound containing oxygen recovered — hydronium ion obtained

Quiz: Structure and Properties

1. What is the electronic configuration of sulfur?

   (A) $1s^2\ 2s^2\ 2p^8\ 3s^2\ 3p^2$

   (B) $1s^2\ 2s^2\ 2p^6\ 3s^2\ 3p^4$

   (C) $1s^2\ 2s^2\ 3s^2\ 2p^8\ 3p^2$

   (D) $1s^2\ 2s^2\ 2p^8\ 3s^2\ 3p^8$

   (E) $1s^2\ 2s^2\ 3s^2\ 2p^8\ 3p^8$

2. sp^2 hybridization will be found for carbon in

   (A) CH_4.

   (B) C_2H_4.

   (C) C_2H_2.

   (D) CH_3OH.

   (E) CH_3OCH_3.

3. Which of the following is responsible for the abnormally high boiling point of water?

   (A) Covalent bonding

   (B) Hydrogen bonding

(C) High polarity

(D) Large dielectric constant

(E) Low molecular weight

4. _______ involves electron transfer from one atom to another?

(A) Covalent bonding

(B) Ionic bonding

(C) Hydrogen bonding

(D) π bonding

(E) Metallic bonding

5. For the molecules listed below, the resultant dipole moments are oriented as (from left to right)

H–O–H   O=C=O   O=S=O

(A) O, $\rightarrow$, $\leftarrow$.

(B) $\uparrow$, O, $\downarrow$.

(C) $\uparrow$, O, $\uparrow$.

(D) $\downarrow$, O, $\uparrow$.

(E) $\uparrow$, $\leftarrow$, $\downarrow$.

6. Which of the following is most soluble in water?

(A) Hexanol

(B) Benzene

(C) Acetic acid

(D) Acetylene

(E) Hexanoic acid

7. Arrange the acids in order of increasing strength.

(A) HCl, O_4, H_2SO_4, H_3PO_4, HClO

(B) HClO, $HClO_4$, H_2SO_4, H_3PO_4

(C) H_3PO_4, H_2SO_4, $HClO_4$, $HClO$

(D) $HClO$, H_3PO_4, H_2SO_4, $HClO_4$

(E) H_3PO_4, H_2SO_4, $HClO$, $HClO_4$

8. Which of the following best describes the diagram below of a molecular orbital?

(A) A nonbonding orbital

(B) A bonding σ orbital

(C) An antibonding σ orbital

(D) A bonding π orbital

(E) An antibonding π orbital

9. Which of the compounds below has dipole moment zero?

(A) CH_4

(B) CH_3Cl

(C) H_2O

(D) NH_3

(E) HF

10. Which of the following bonds (• • •) is the least polar?

(A) B • • • Cl

(B) H • • • I

(C) P • • • Br

(D) C • • • Cl

(E) C • • • I

ANSWER KEY

1.	(B)	6.	(C)
2.	(B)	7.	(D)
3.	(B)	8.	(E)
4.	(B)	9.	(A)
5.	(B)	10.	(E)

CHAPTER 2

Alkanes

Structural Formula: C_nH_{2n+2}

The simplest member of the alkane family is methane (CH_4), which is written as:

```
     H                        H
     |                        ..
 H—C—H          or        H : C : H
     |                        ..
     H                        H
```

2.1 Nomenclature (IUPAC System)

A) Select the longest continuous carbon chain for the parent name.

Example

```
CH3—CH2—CH—CH2—CH2—CH3
         |
         CH3
```

The parent name is hexane.

B) Number the carbons in the chain, from either end, such that the substituents are given the lowest numbers possible.

Example

$$\overset{1}{C}H_3-\overset{2}{C}H_2-\overset{3}{C}H-\overset{4}{C}H_2-\overset{5}{C}H_2-\overset{6}{C}H_3$$
$$\quad\;\; | $$
$$CH_3$$

C) The substituents are assigned the number of the carbon to which they are attached. In the preceding example, the substituent is assigned the number 3.

D) The name of the compound is now composed of the name of the parent chain preceded by the name and the number of the substituents, arranged in alphabetic order. For the same example, the name is thus 3-methylhexane.

E) If a substituent occurs more than once in the molecule, the prefixes, "di-," "tri-," "tetra-," etc., are used to indicate how many times it occurs.

F) If a substituent occurs twice on the same carbon, the number of the substituent is repeated.

Example

$$CH_3-CH_2-\underset{CH_3}{\overset{CH_3}{C}}-CH_2-CH_3$$ 3,3-dimethylpentane

$$CH_3-CH_2-CH_2-\underset{CH_3-CH-CH_3}{CH}-\underset{CH_3}{CH}-\underset{CH_2-CH_3}{\overset{CH_2-CH_3}{C}}-CH_2-CH_3$$

3, 3–diethyl–5–isopropyl–4–methyloctane

Problem Solving Examples:

Name the following alkanes.

(a) CH_4

(b) CH_3CH_3

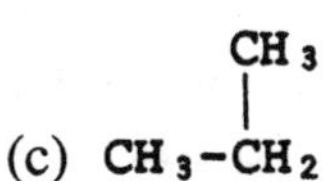

(d) $CH_3CH_2CH_2CH_3$

Four steps can be followed in naming alkanes.

Step I: In naming open-chain alkanes, first find the longest chain of carbon atoms.

Step II: Write down the parent name.

Step III: Identify any side chains.

Step IV: Number these side chains and add their names and locations as a prefix to the name of the parent compound.

These steps are illustrated in the naming of the four compounds above.

(a)

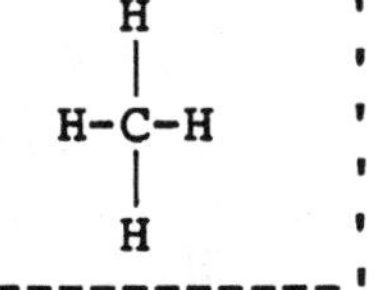

This compound contains only one carbon atom and is the simplest of all the alkanes. It is called methane.

(b)

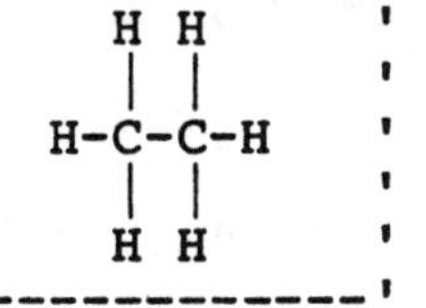

This is a two-carbon alkane. A chain of two carbon atoms is given the root "eth." The parent names of alkanes are formed by adding the suffix "ane" to the root name of the longest carbon chain. This compound is called ethane.

(c)

```
           H
           |
    H    H-C-H
    |      |
  H-C -----C-H
    |      |
    H      H
```

The longest chain in this compound is three carbons long. It is not significant that the chain is bent. The root used in naming three-carbon chains is "prop." The name of this compound is propane.

(d)

```
    H H H H
    | | | |
  H-C-C-C-C-H
    | | | |
    H H H H
```

The longest chain here is four carbons long. The root used in naming four-carbon chains is "but." The name of this compound is butane.

Name the following compounds:

```
(a)       CH3
          |
    CH3-CH-CH2-CH2-CH-CH3
                   |
                   CH2CH3

(b) CH3CH2CH-CH2-CH-CH2-CH2-CH3
           |       |
           CH3     CH2CH3

(c)         CH2CH2CH3
            |
    CH3CH2CH-CH2CH2CHCH3
                   |
                   CH3
```

In brief, the rules for naming organic compounds are as follows:

(1) determine and identify the longest continuous carbon chain,
(2) locate and name any side chains, and

(3) add the names and positions of the side chains as a prefix to the name of the parent compound.

After looking at the structures above, it is seen that all of the compounds are saturated hydrocarbons or alkanes. They all have the formula C_nH_{2n+2}, where *n* is the number of carbon atoms. Another characteristic of alkanes is that the compounds contain only single or sigma bonds.

(a)
```
         CH3
         |
 1CH3 2CH - 3CH2 4CH2 5CH-CH3
                       |
                      6CH2 7CH3
```

The longest continuous chain contains seven carbons. These carbons have been numbered in the accompanying diagram.

The root of the name of a seven-carbon chain is "hept-."

The carbons are numbered from lowest to highest beginning with the end of the chain that contains the side chain of greatest molecular weight closest to the end of the chain. The substituents on this chain are two CH_3– groups placed one at C2 and one at C5. The CH_3– side chain has one carbon and is called methyl. The name of this compound is 2,5-dimethylheptane.

(b)
```
 1CH3 2CH2 3CH- 4CH2 5CH -6CH2 7CH2 8CH3
            |         |
           CH3       CH2CH3
```

As shown, the longest carbon chain contains eight carbons. The root used in naming eight-carbon chains is "oct–." As the accompanying structure shows, the numbering proceeds from the right end of the structure toward the left. The direction is so chosen as to give the lowest numbers possible to the side chains. If two or more side chains of different nature are present, they are cited in alphabetical order. The name of this compound is 5-ethyl-3-methyloctane.

(c)

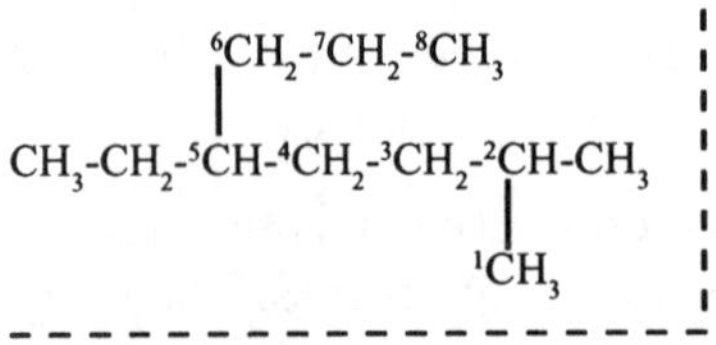

$$\begin{array}{l} \qquad\qquad\quad {}^{6}CH_2\text{-}{}^{7}CH_2\text{-}{}^{8}CH_3 \\ \qquad\qquad\quad | \\ CH_3\text{-}CH_2\text{-}{}^{5}CH\text{-}{}^{4}CH_2\text{-}{}^{3}CH_2\text{-}{}^{2}CH\text{-}CH_3 \\ \qquad\qquad\qquad\qquad\qquad\quad | \\ \qquad\qquad\qquad\qquad\qquad {}^{1}CH_3 \end{array}$$

The longest carbon chain in this compound contains eight carbons. The lowest numbering scheme for the substituents is a methyl group on C2 and an ethyl group on C5, as shown in the accompanying structure. If two or more side chains of different nature are present, they are cited in alphabetical order. The name of this compound is 5-ethyl-2-methyloctane.

2.2 Physical Properties of Alkanes

Alkanes consist of nonpolar or very weak polar molecules, which are attracted to each other in the liquid or solid phase by weak van der Waals forces.

The boiling point, melting point, density, and viscosity increase as the length of the carbon chain increases.

In general, branched chain alkanes exhibit lower boiling points than corresponding straight-chain alkanes.

The alkanes are soluble in all nonpolar or weak polar solvents, such as benzene, chloroform, ether, and carbon tetrachloride, because "like dissolves like."

The melting point of an alkane is not only dependent on the size of the molecule, but also on the ease with which the molecule fits into a crystal lattice.

Problem Solving Example:

Q Isobutane is thermodynamically more stable than butane. Which has the lower boiling point? Is there any relationship between thermodynamic stability and boiling point? Would you expect such a relationship between thermodynamic stability and melting point?

A Thermodynamic stability refers to the intramolecular forces (van der Waals forces) acting within the molecular structure of a compound. Thus, benzene

which has a low ground state energy, is considered to be thermodynamically stable. An enormous amount of energy is required for the degradation of this compound. However, the boiling point is related to the intermolecular forces in a given compound. When the temperature is raised, thus increasing the kinetic energy, the molecular collisions are greatly enhanced and the distance between molecules is widened. The transition from the liquid to the vapor phase occurs and the relatively cohesive liquid is transformed to wandering molecules in the gaseous phase.

```
[        CH3         ]
[         |          ]
[   H3C-C—CH3        ]
[         |          ]
[         H          ]
```

Isobutane (shown above) is expected to have a lower boiling point (b.p. = – 12°C) than the straight chain isomer (b.p. = 0°C). The boiling points of alkanes rise as the molecules get larger. Branching decreases the boiling point because the shape of the molecule approaches that of a sphere, thus decreasing the surface area. This results in the weakening of the intermolecular forces, which are overcome at a lower temperature.

Considering the above descriptions of stability and boiling points, there is absolutely no correlation between the two.

2.3 Rotational Structures

The atoms in alkane molecules are joined by σ-bonds, and the electron distribution is cylindrically symmetrical about a line joining the atomic nuclei.

There exists an infinite number of free rotations about the carbon-carbon single bonds of the alkanes. Each rotation results in the rearrangement of the atoms of the molecules and is called a conformation.

Take for example ethane, with its three-dimensional representation shown in Figure 2.1. The frontal carbon atom is represented by a point and the atom behind it, by a circle; such representation is called a Newman projection.

If the frontal carbon atom in (a) is rotated 60° and the other is held fixed, a second Newman projection is achieved, as shown in (b). These two arrangements of the atoms of the molecule are referred to respectively as the staggered and eclipsed conformations of ethane.

(a) staggered conformation (b) eclipsed conformation

Figure 2.1 Newman Projections of Ethane

Problem Solving Example:

Draw perspective formulas and Newman projections for the eclipsed and staggered forms of propane.

A Propane (C_3H_8) is a member of the paraffins (saturated hydro carbons). The eclipsed and staggered forms of a molecule are different conformations of that molecule. Conformations of a molecule are structures of that molecule that differ only in the torsional angle of carbon-carbon single bonds. Since perspective formulas are three dimensional, we can represent them on paper by having a dashed bond represent a projection away from the reader, the heavy wedge bond projecting towards the reader, while a normal bond would represent the plane of the paper.

In the eclipsed form of propane, there is more steric hindrance than in the staggered form. There is also a greater torsional strain in the eclipsed form. As a result, the eclipsed form is expected to be higher in energy than the staggered form. Therefore, the staggered conformation is more stable. The perspective formulas and Newman projections are sketched below:

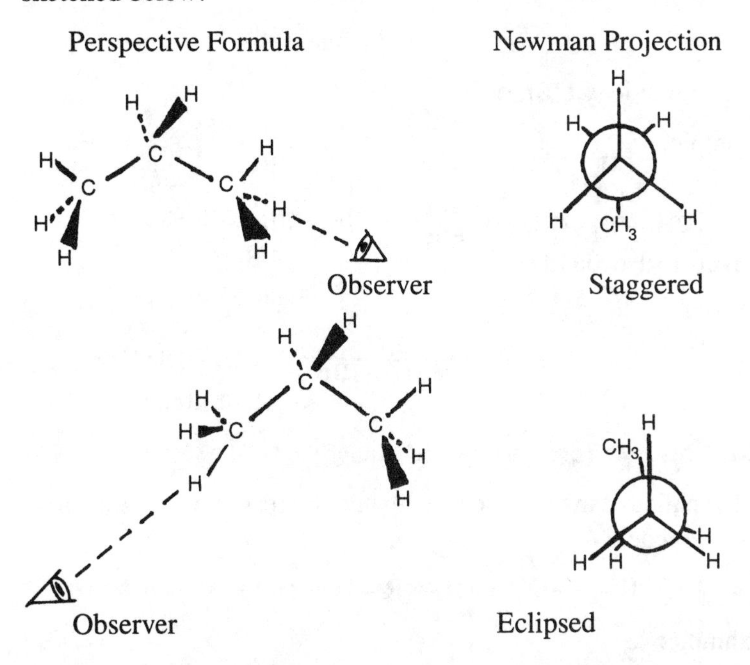

2.4 Preparation

Hydrogenation of Alkenes

Alkenes are converted into alkanes of the same carbon skeleton by the addition of hydrogen atoms to the double bond.

$$C_nH_{2n} \xrightarrow[\text{or Ni}]{H_2 + Pt} C_nH_{2n+2}$$

Hydrolysis of Grignard Reagents

Grignard reagents are organometallic compounds, RMgX (i.e., compounds that contain carbon-metal covalent bonds). They are very reactive and can be prepared by reaction of magnesium metal, Mg, and an alkyl halide, R – X, in dry ether solvent.

$$R{-}X + Mg \xrightarrow{\text{dry ether}} RMgX \xrightarrow{H_2O} R{-}H$$

alkyl halide Grignard reagent alkane

where: X = halide (Cl, Br, or I).

Example:

$$CH_3{-}\underset{\displaystyle CH_3}{\overset{|}{CH}}{-}CH_2{-}Br \xrightarrow[\text{ether}]{Mg} CH_3{-}\underset{\displaystyle CH_3}{\overset{|}{CH}}{-}CH_2{-}MgBr$$

isobutyl bromide

$$\xrightarrow[-Mg(OH)Br]{+H_2O} CH_3{-}\underset{\displaystyle CH_3}{\overset{|}{CH}}{-}CH_3$$

isobutane

Reduction by Metal and Acid

Alkyl halides can be reduced to alkanes by reaction with zinc metal and a mineral acid.

$$R - X + Zn + H^+ \rightarrow R - H + Zn^{2+} + X^-$$

Example:

$$CH_3{-}\underset{Br}{CH}{-}\underset{CH_3}{CH}{-}CH_2{-}CH_3 \xrightarrow[0°C]{Zn, HI} CH_3{-}CH_2{-}\underset{CH_3}{CH}{-}CH_2{-}CH_3$$

2–bromo–3-methylpentane 3–methylpentane

Reduction with Alkali Metal Hydrides

The strong reducing agents, lithium aluminum hydride, $LiAlH_4$, and sodium borohydride, $NaBH_4$, readily reduce alkyl halides to alkanes.

$$R{-}X + LiAlH_4 \xrightarrow{\text{dry ether}} R{-}H$$

Examples:

$$CH_3{-}\underset{CH_3{-}CH{-}CH_3}{CH}{-}CH_2{-}Br + LiAlH_4 \xrightarrow[\text{ether}]{\text{dry}} 4CH_3{-}\underset{CH_3{-}CH{-}CH_3}{CH}{-}CH_3 + LiBr + AlBr_3$$

1–bromo–2, 3–dimethylbutane 2, 3–dimethylbutane

The Wurtz Reaction

When two moles of an alkyl halide and two moles of metallic sodium are reacted, an alkyl sodium is formed which reacts with a second alkyl halide to form the alkane:

$$2R - X + 2Na \rightarrow R - Na + NaX$$

$$\frac{R - Na + X - R \rightarrow R - R + NaX}{2R - X + 2Na \rightarrow R - R + 2NaX}$$

The reaction between sodium and two different alkyl halides yields a mixture of three different alkanes.

$$3RX + 3R'X + 6Na \rightarrow R - R + R' - R' + R' - R + 6NaX$$

The Kolbe Synthesis

The electrolysis of sodium, potassium, or calcium salts of carboxylic acids yields alkanes.

$$2CH_3{-}CH_2{-}CO\bar{O}\,Na^+ + 2H_2O \xrightarrow{\text{electrolysis}} CH_3{-}CH_2{-}CH_2{-}CH_3$$

sodium propanoate — n-butane (on anode)

$2CO_2 + 2NaOH + H_2$

on cathode

Problem Solving Examples:

Write equations for the preparation of n-butane from:

(a) n-butyl bromide
(b) sec-butyl bromide
(c) ethyl chloride
(d) 1-butene, $CH_3CH_2CH = CH_2$
(e) 2-butene, $CH_3CH = CHCH_3$

The structure of n-butane (C_4H_{10}) may be written as:

```
      H  H
      |  |
H3C-C-C-CH3
      |  |
      H  H
```

(a) To prepare n-butane from n-butyl bromide ($CH_3CH_2CH_2CH_2Br$), first take note that both compounds have the same carbon content. This means the reactions to remove the bromide must not alter the carbon skeleton. To replace the bromine atom with a hydrogen atom without altering the carbon skeleton, employ a reaction that results in the reduction of the alkyl halide. Two such reactions are the hydrolysis of Grignard reagent and the reduction by metal and acid.

When a solution of alkyl halide is placed in dry ethyl ether with metallic magnesium, a vigorous reaction takes place—a Grignard reagent is formed. The Grignard reagent has the general formula RMgX, where R is an alkyl or aryl group and X is a halogen atom. The Grignard reagent is extremely reactive. It reacts with numerous inorganic and organic compounds. If water is added, hydrolysis occurs to produce R – H, a hydrocarbon of the same carbon skeleton. The hydrolysis may be viewed as the reaction of a salt (the Grignard reagent) with an acid (H_2O) to produce a weaker acid (R – H). This reaction, which can produce the n-butane desired, is illustrated in the following reaction sequence:

$$\underset{\text{(n-butyl bromide)}}{CH_3CH_2CH_2CH_2Br} \xrightarrow[\text{Ether (Dry)}]{Mg} \underset{\text{(n-butylmagnesium bromide)}}{CH_3CH_2CH_2CH_2MgBr} \xrightarrow{H_2O}$$

$$\underset{\text{(n-butane)}}{CH_3CH_2CH_2CH_3} + Mg(OH)Br$$

Reduction of n-butyl bromide with metal and acid will accomplish the same thing as shown:

$$CH_3CH_2CH_2CH_2Br + Zn + H+ \rightarrow CH_3CH_2CH_2CH_3 + ZnBr_2$$

(b) Sec-butyl bromide may be written as:

$$\underset{\underset{Br}{|}}{CH_3CH_2CHCH_3}$$

Here, again, hydrolysis of Grignard reagent or reduction by metal and acid gives the desired product, n-butane.

$$CH_3CH_2\underset{\underset{Br}{|}}{C}HCH_3 \xrightarrow[\text{(dry)}]{\text{Mg, Ether}} CH_3CH_2\underset{\underset{MgBr}{|}}{C}HCH_3 \xrightarrow{H_2O} CH_3CH_2CH_2CH_3$$

$$CH_3CH_2\underset{\underset{Br}{|}}{C}HCH_3 + Zn + H^+ \longrightarrow CH_3CH_2CH_2CH_3 + ZnBr_2$$

(c) Ethyl chloride possesses only a two-carbon backbone: CH_3CH_2Cl. To produce n-butane, which has four carbons, one must use a reaction that will add two carbons. To manufacture an alkane of higher carbon number than the starting material will require the formation of carbon-carbon bonds. This is most directly accomplished by the coupling together of two alkyl groups. An excellent method of doing this was developed by Corey and House. It is the coupling of alkyl halides with organometallic compounds. Coupling takes place between a lithium dialkylcopper, R_2CuLi, and an alkyl halide, R'X (R" may be different or the same as R). The general reaction may be illustrated as shown:

$$R_2CuLi + R'X \rightarrow \underset{\text{(Alkane)}}{R - R'} + RCu + LiX$$

The lithium dialkylcopper is prepared by adding lithium to the alkyl halide, which may be primary, secondary, or tertiary. After this, one adds cuprous halide, CuX. Overall,

$$RX \xrightarrow{Li} \underset{\text{(Alkyl lithium)}}{RLi} \xrightarrow{CuX} \underset{\text{(Lithium dialkylcopper)}}{R_2CuLi}$$

To alkyl halide (R'X) that will be added to R_2CuLi (to produce R – R') should be primary to obtain good yields. The reaction sequence that generates n-butane from ethyl chloride can now be written:

$$CH_3CH_2Cl \xrightarrow{Li} CH_3CH_2Li \xrightarrow{CuX} (CH_3CH_2)_2CuLi$$

$$(CH_3CH_2)_2CuLi + CH_3CH_2Cl \rightarrow CH_3CH_2CH_2CH_3 + CH_3CH_2Cu + LiCl$$

(d) and (e) Alkenes (compounds with double bonds) such as

1-butene and 2-butene can be readily converted to the corresponding alkane with a hydrogenation reaction. In this reaction, hydrogen gas is passed over the alkene in the presence of Pt, Pd, or Ni, which act as catalysts. The hydrogen adds across the double bond to produce an alkane of the same carbon number. Hydrogenation of 1-butene and 2-butene directly produces the desired product as shown:

$$CH_3CH_2CH=CH_2 + H_2 \xrightarrow{\text{Pt, Pd, or Ni}} CH_3CH_2CH_2CH_3$$
(1 – butene)

$$CH_3CH=CHCH_3 + H_2 \xrightarrow{\text{Pt, Pd, or Ni}} CH_3CH_2CH_2CH_3$$

The following projected synthesis for n-butane is not very efficient. Why?

$$CH_3CH_2CH_2Br + CH_3Br + 2Na \rightarrow CH_3(CH_2)_2CH_3 + 2NaBr$$

This reaction is an example of the Wurtz coupling reaction. The general equation is

$$2R-X + 2Na \rightarrow R-R + 2NaX$$

where R represents an alkyl group and X a halide (Cl, Br, or I).

In the above reaction there is both n-propylbromide and methylbromide in the mixture. According to the equation for the Wurtz reaction, these two compounds can couple to form n-butane. But, at the same time free methyl groups can combine with each other as can free propyl groups, forming ethane and hexane. This leads to a decrease in the production of n-butane. A more efficient reaction mixture would be just ethylbromide and Na. Here, the only coupled product would be n-butane

$$CH_3-CH_2-Br + 2Na \rightarrow CH_3(CH_2)_2CH_3 + 2NaBr.$$

2.5 Reactions of Alkanes

Oxidation of Alkanes

The combustion of alkanes gives carbon dioxide and water, with the evolution of large quantities of heat.

$$CH_4 + 2O_2 \xrightarrow{\text{flame}} CO_2 + 2H_2O$$

$$2CH_3CH_3 + 7O_2 \xrightarrow{\text{flame}} 4CO_2 + 6H_2O$$

Formation of Alkyl Hydroperoxides

Good yields are obtained from preparation with tertiary carbons.

Example:

$$CH_3-\underset{CH_3}{\overset{CH_3}{\underset{|}{\overset{|}{C}}}}-H + O_2 \xrightarrow{140°C} CH_3-\underset{CH_3}{\overset{CH_3}{\underset{|}{\overset{|}{C}}}}-O-O-H$$

isobutane　　　　tert-butylhydroperoxide

Halogenation of Alkanes

In the presence of heat or ultraviolet light, either chlorine or bromine reacts with an alkane to produce an alkyl halide and a hydrohalogen acid.

Example: Chlorination of methane

$$H-\underset{H}{\overset{H}{\underset{|}{\overset{|}{C}}}}-H + Cl-Cl \xrightarrow[\text{or peroxides}]{\text{light, heat}} H-\underset{H}{\overset{H}{\underset{|}{\overset{|}{C}}}}-Cl + H-Cl$$

methane　　　　chloromethane

This is an example of a substitution reaction. On further halogenation, all hydrogen atoms in the methane molecule will be replaced by halogen atoms.

The reactivity of a hydrogen atom during the halogenation process depends chiefly on its class and not on the alkane to which it is attached. The relative ease with which hydrogen atoms can be abstracted is:

$$3° > 2° > 1° > CH_4$$

Problem Solving Examples:

Q Why would the bromination of ethane to ethyl bromide be a more efficient synthesis than the bromination of hexane to 1-bromohexane?

A There is only one possible monobrominated product formed from ethane, that is, ethyl bromide. This is true because the two carbons making up the ethane molecule are equivalent and the bromine radical will attack either one with equal frequency. This makes the synthesis of ethyl bromide from ethane an extremely efficient process.

This is not true of the production of 1-bromohexane from hexane. The six carbon atoms comprising hexane are not equivalent and will not be attacked by the bromine radical with equal frequency. In hexane there are two sets of equivalent carbons. One set is composed of the two carbons at the ends of the chain. These carbons are referred to as primary carbons because they are bound to only one other carbon atom. The four carbons in the center of the chain comprise the other set. These carbons are referred to as secondary carbons because they are each bound to two other carbon atoms. Secondary carbons are more reactive than primary carbons and will bind the bromine radical more readily. This will cause the production of more 2- and 3-bromohexane than 1-bromohexane. As a synthesis for 1-bromohexane, monobromination of hexane is a rather inefficient procedure.

Predict the expected products of the following substitution reactions. What is the reason for your answer?

(a) $CH_3-CH_2-CH_3 + Cl_2 \xrightarrow[\text{or peroxides}]{\text{light, heat}}$

(b) $CH_3-\underset{H}{\overset{CH_3}{C}}-CH_3 + Cl_2 \longrightarrow$

(c) $CH_3-\underset{CH_3}{\overset{CH_3}{C}}-CH_3 + Cl_2 \longrightarrow$

(d) $C_6H_{11}-CH_3 + Cl_2 \longrightarrow$

(a) $CH_3-CH_2-CH_3 + Cl_2 \longrightarrow CH_3\underset{Cl}{CH}-CH_3$

(b) $CH_3-\underset{H}{\overset{CH_3}{C}}-CH_3 + Cl_2 \longrightarrow CH_3-\underset{Cl}{\overset{CH_3}{C}}-CH_3$

(c) $CH_3-\underset{CH_3}{\overset{CH_3}{C}}-CH_3 + Cl_2 \longrightarrow CH_3-\underset{CH_3}{\overset{CH_3}{C}}-CH_2Cl$

(d) $C_6H_{11}-CH_3 + Cl_2 \longrightarrow$ (1-Cl, 1-CH_3 ring) + $C_6H_{11}-CH_2Cl$ + (1-CH_3, 2-Cl ring) + (1-CH_3, 3-Cl ring) + (1-CH_3, 4-Cl ring)

The reactivity of a hydrogen atom during halogenation depends chiefly on its class. The relative ease of hydrogen abstraction is $3^\circ > 2^\circ > 1^\circ > CH_4$.

2.6 Structural Isomerism

There are two types of butanes—normal butane and iso-butane. They have the same molecular formula, C_4H_{10}, but have different structures. n-butane is a straight-chain hydrocarbon, whereas iso-butane is a branched-chain hydrocarbon. n-butane and iso-butane are structural isomers and differ in their physical and chemical properties.

```
                                       H  H  H  H
                                       |  |  |  |
C4H10 ≡ CH3CH2CH2CH3  ≡             H–C––C––C––C–H     n-butane
                                       |  |  |  |        (straight chained)
                                       H  H  H  H
```

$C_4H_{10} \equiv CH_3CH_2CH_2CH_3$ — n-butane (straight chained)

```
                                       H  H  H
                                       |  |  |
C4H10 ≡ CH3CH(CH3)CH3 ≡             H–C––C––C–H        isobutane
                                       |  |  |           (branched)
                                       H  |  H
                                          |
                                        H–C–H
                                          |
                                          H
```

$C_4H_{10} \equiv CH_3CH(CH_3)CH_3$ — isobutane (branched)

In higher homologs of the alkane family, the number of isomers increases exponentially.

Problem Solving Example:

Draw all the isomers of C_5H_{12}.

A systematic approach for drawing all the structures of a compound is necessary. First, start with the straight-chain isomer. Next, draw the possible structures using a straight-chain of carbon length one less than the first isomer. Repeating this second step will lead to the formation of all of the isomers. This method will be illustrated using pentane (C_5H_{12}). The straight-chain isomer is

```
  H H H H H
  | | | | |
H-C-C-C-C-C-H
  | | | | |
  H H H H H
```

Straight-chain isomers are designated by adding the prefix n to the name of the compound. This isomer is called n-pentane.

The next shortest chain contains four carbons. The fifth carbon appears as a methyl group at C2. There is only this one isomer with a chain length of four.

```
                  H
                  |
H H             H-C-H      H
| |               |        |
H-C-C ----------- C ------ C-H
| |               |        |
H H               H        H
```

This structure is called 2-methylbutane.

When the chain is reduced to three carbons, one further isomer can be drawn.

```
          H
          |
        H-C-H
    H     |     H
    |     |     |
  H-C  -  C  -  C-H
    |     |     |
    H     |     H
        H-C-H
          |
          H
```

The name of this structure is 2,2-dimethyl propane or neopentane.

When the carbon chain is reduced to two carbons, the isomer 2-methylbutane is produced again. It is exactly the same isomer as the one with four carbon atoms in the straight chain.

2.7 Free Radical Reactions

Chlorination by Sulfuryl Chloride

$$R-H + SO_2Cl_2 \xrightarrow[40-80°C]{\text{light or peroxide}} R-Cl + SO_2 + HCl$$

Sulfochlorination

Alkanes are sulfo-chlorinated by sulfuryl chloride, in the presence of a base to form alkane sulfonylchlorides.

$$R-H + SO_2Cl_2 \xrightarrow{\text{base}} R-SO_2Cl + HCl$$

Nitration

In the vapor phase, alkanes can be nitrated, at high temperature, with HNO_3 or N_2O_4 to form a mixture of nitroalkanes.

$$R\text{–}H + HNO_3 \xrightarrow{> 400^\circ C} R\text{–}NO_2 + H_2O$$

Problem Solving Example:

Q Predict the proportions of isomeric products from chlorination at room temperature of: (a) propane; (b) isobutane; (c) 2,3 dimethylbutane; (d) n-pentane (Note: There are three isomeric products); and (e) isopentane.

A The answer to this problem involves a consideration of orientation, the factors that determine where in a molecule reaction is most likely to occur. Orientation is determined by the relative rates of competing reactions.

Suppose one compared the rate of abstraction of primary hydrogens with the rate of abstraction of secondary hydrogens. What are the factors that determine the rates of these two reactions?

(1) Collision frequency. Since both reactions involve collision of the same particles, a specific alkene and a halogen, this must be the same for the two reactions. (2) Probability factor. If a primary hydrogen is to be abstracted, the alkane must be so oriented at the time of collision that the halogen atom strikes a primary hydrogen. Likewise, if a secondary hydrogen is to be abstracted, the alkane must be so oriented that the halogen atom strikes a secondary hydrogen. Now, if the alkane has, say, six primary hydrogens and only two secondary hydrogens, it could be estimated that the probability factor favors abstraction of primary hydrogens by the ratio of 6:2, or 3:1. (3) Energy of activation (E_{act}) is more for abstraction of a primary hydrogen than for abstraction of a secondary hydrogen. In fact, the E_{act} is greater for abstraction of a secondary hydrogen than for abstraction of a tertiary hydrogen. Hence, abstraction of hydrogens is found to follow the sequence $3^\circ > 2^\circ > 1^\circ$ in reactivity. In chlorination at room temperature, the relative rates per hydrogen atom are 5.0:3.8:1.0.

With this information, the proportions of isomeric products from chlorination can be predicted.

(a) Propane's isomeric products (from chlorination):

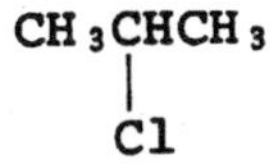

$$\underset{\text{(isopropyl chloride)}}{CH_3\underset{\substack{|\\ Cl}}{C}HCH_3} \qquad \underset{\text{(n-propyl chloride)}}{CH_3CH_2CH_2Cl}$$

To figure out the relative amounts of these products, remember that the probability and reactivity (E_{act}) of the hydrogens must be examined. In n-propyl chloride, the chlorine halogen abstracted a primary hydrogen (1°H) of relative rate 1, whereas in isopropyl chloride the halogen abstracted a secondary hydrogen (2°H) of relative rate 3.8. The calculation of isomeric proportion is as follows:

$$\frac{\text{n-PrCl}}{\text{i-PrCl}} = \frac{\text{no. of 1° H}}{\text{no. of 2° H}} \times \frac{\text{reactivity of 1° H}}{\text{reactivity of 2° H}} = \frac{6}{2} \times \frac{1.0}{3.8} = \frac{6.0}{7.6}$$

Hence, the percentage of 1° isomeric product (that is, n-propyl chloride) is equal to $\frac{6.0}{6.0 + 7.6} \times 100 = 44\%$. Consequently, percentage of 2° isomeric product (that is, isopropyl chloride) is 56%, which is $\frac{7.6}{6.0 + 7.6} \times 100$.

(b) Isobutane's isomeric products:

$$\underset{\text{(tert-butyl chloride)}}{CH_3-\overset{\substack{CH_3\\ |}}{\underset{\substack{|\\ Cl}}{C}}-CH_3} \qquad \underset{\text{(iso-butyl chloride)}}{CH_3-\overset{\substack{CH_3\\ |}}{\underset{\substack{|\\ H}}{C}}-CH_2Cl}$$

The method of calculation of their proportion is the same as in (a).

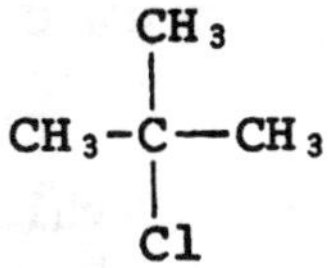

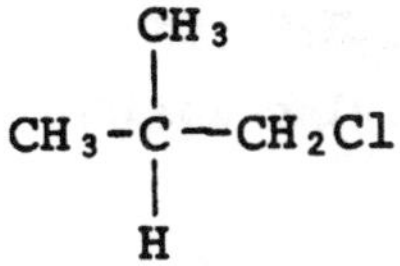

$$\frac{\text{tert-butyl chloride}}{\text{iso-butyl chloride}} = \frac{\text{no. of 3° H}}{\text{no. of 1° H}} \times \frac{\text{reactivity of 3° H}}{\text{reactivity of 1° H}}$$

$$= \frac{1}{9} \times \frac{5}{1} = \frac{5}{9}$$

Therefore, the percentage of t-butyl-chloride =

$$\frac{5}{5 + 9} \times 100 = 36\%$$

The percentage of iso-butyl chloride must be

$$\frac{9}{9+5} \times 100 = 64\%$$

(c) 2,3-dimethylbutane's isomeric products:

$$H_3C-\underset{Cl}{\overset{CH_3}{C}}-\underset{H}{\overset{CH_3}{C}}-CH_3$$

(2-chloro-2,3-dimethyl butane)

$$H_3C-\underset{H}{\overset{CH_3}{C}}-\underset{H}{\overset{CH_3}{C}}-CH_2Cl$$

(1-chloro-2,3-dimethyl-butane)

$$\frac{\text{2-chloro}}{\text{1-chloro}} = \frac{\text{no. of } 3° \text{ H}}{\text{no. of } 1° \text{ H}} \times \frac{\text{reactivity of } 3° \text{ H}}{\text{reactivity of } 1° \text{ H}}$$

$$= \frac{2}{12} \times \frac{5}{1} = \frac{10}{12} = \frac{5}{6}$$

Consequently, $\frac{5}{5+6} \times 100 = 45\%$ is the percentage of 2-chloro-2,3-dimethylbutane, while $\frac{6}{6+5} \times 100 = 55\%$ is the amount of 1-chloro-2,3-dimethyl-butane.

(d) n-pentane's isomeric products:

$CH_3CH_2CH_2CH_2CH_2Cl$ (n-pentyl chloride)

$CH_3CH_2CH_2\underset{Cl}{CH}CH_3$ (2-chloropentane)

$CH_3CH_2\underset{Cl}{CH}CH_2CH_3$ (3-chloropentane)

n-pentyl chloride:	six primary hydrogens with reactivity of 1.0, so that $6 \times 1.0 = 6$.
2-chloropentane:	four secondary hydrogens with reactivity of 3.8, so that $4 \times 3.8 = 15.2$.
3-chloropentane:	two secondary hydrogens with reactivity of 3.8, so that $2 \times 3.8 = 7.6$.

The total number × reactivity = 6.0 + 15.2 + 7.6 = 28.8. Hence, the percentage of n-pentyl chloride is $\frac{6}{28.8} \times 100 = 21\%$.

2-chloropentane: $\frac{15.2}{28.8} \times 100 = 53\%$

3-chloropentane: $\frac{7.6}{28.8} \times 100 = 26\%$.

(e) Isopentane's isomeric products:

$$CH_3-\underset{\underset{H}{|}}{\overset{\overset{CH_3}{|}}{C}}-CH_2CH_2Cl$$

(1-chloro-3-methylbutane)

$$CH_3-\underset{\underset{Cl}{|}}{\overset{\overset{CH_3}{|}}{C}}-CH_2CH_3$$

(2-chloro-2-methylbutane)

$$CH_2Cl-\underset{\underset{H}{|}}{\overset{\overset{CH_3}{|}}{C}}-CH_2CH_3$$

(1-chloro-2-methyl butane)

$$CH_3-\underset{\underset{H}{|}}{\overset{\overset{CH_3}{|}}{C}}-\underset{\underset{Cl}{|}}{C}HCH_3$$

(3-chloro-2-methylbutane)

1-chloro-3-methylbutane: three primary hydrogens with reactivity of 1.0.3 × 1 = 3.

2-chloro-2-methylbutane: one tertiary hydrogen with reactivity of 5.0.5 × 1 = 5.

1-chloro-2-methylbutane: six primary hydrogens with reactivity of 1.0.6 × 1 = 6.

3-chloro-2-methylbutane: two secondary hydrogens with reactivity of 3.8.2 × 3.8 = 7.6.

The total number × reactivity = 3 + 5 + 6 + 7.6 = 21.6.

The percentages for the four isomeric compounds are as follows:

1-chloro-3-methylbutane: $\frac{3}{21.6} \times 100 = 14\%$

2-chloro-2-methylbutane: $\frac{5}{21.6} \times 100 = 23\%$

1-chloro-2-methylbutane: $\frac{6}{21.6} \times 100 = 28\%$

3-chloro-2-methylbutane: $\frac{7.6}{21.6} \times 100 = 35\%$

Quiz: Alkanes

1. Isomers differ in

   (A) the number of neutrons in their nuclei.

   (B) their atomic compositions.

   (C) their molecular weights.

   (D) their molecular structures.

   (E) None of the above.

2. Which of the following structures represents 1,1-dibromoethane?

   (A)
```
     Br  H
     |   |
Br - C — C - H
     |   |
     H   H
```

   (B)
```
    Br  Br
    |   |
H - C — C - H
    |   |
    H   H
```

   (C) $CH_3 - CH_2 - BR - CH_2 - CH_3$

   (D) $BR - C \equiv C - BR$

```
(E) Br      H
      \    /
       C=C
      /    \
    Br      H
```

3. What type of formula of ethane is depicted below?

```
          H
     H    |    H
       \  |  /
        ( | )
       /     \
    H     |     H
          H
```

(A) Fisher projection formula

(B) Newman projection formula

(C) Lewis projection formula

(D) Kekulé projection formula

(E) Pauling projections formula

4. Examine the structures below, then indicate which has the lowest boiling point.

```
(A) C – C – C – C – C

(B) C-C-C-C
        |
        C

        C
        |
(C)   C-C-C
        |
        C

(D) C – C – C – C – C – C

(E) C-C-C-C
      | |
      C C
```

5. Which of the following reacts with hydrogen and nickel to form propane?

   (A) $CH_3CH = CH_2$

   (B) $CH_2CH_2CH_2OH$

   (C) $CH_3\,\underset{\displaystyle |\atop\displaystyle CH_3}{C}HCH_3$

   (D) $CH_3\,\underset{\displaystyle |\atop\displaystyle OH}{C}H — CH_3$

   (E) $CH_3CH_2CH = CH_2$

6. The reduction of 2–butyne to n–butane in the laboratory involves

   (A) the use of an organometallic catalyst.

   (B) the treatment of 2–butyne with hydrogen in the presence of a nickel catalyst.

   (C) the use of an oxidizing agent such as $Cr_2O_4^{2-}$ in the presence of hydrogen.

   (D) the use of a strong base such as KOH, along with sodamide ($NaNH_2$).

   (E) heating 2–butyne, in the presence of the Al_2O_3 catalyst over a stream of hydrogen gas.

7. The following reaction is an example of a _____.

$$2CH_2 - CH_2 - CH_2 - COO - Na^+ + 2H_2O \rightarrow$$
$$CH_3 - CH_2 - CH_2 - CH_2 - CH_2 - CH_3 + 2CO_2 + 2NaOH + H_2$$

   (A) Wurtz reaction. (B) Kolbe synthesis.

   (C) reduction. (D) hydrogenation.

   (E) hydrolysis.

8. Which one of the following is not used in the free radical reactions of alkanes?

(A) SO_2Cl_2 (B) HNO_3

(C) Base (D) Peroxide

(E) Grignard reagent

9. During halogenation of alkanes, the relative ease with which hydrogen atoms can be abstracted is

(A) $CH_4 > 1° > 2° > 3°$.

(B) $1° > CH_4 > 2° > 3°$.

(C) $1° > 2° > CH_4 > 3°$.

(D) $1° > 2° > 3° > CH_4$.

(E) $3° > 2° > 1° > CH_4$.

10. Alkanes are soluble in all of the following EXCEPT

(A) Chloroform.

(B) ether.

(C) water.

(D) benzene.

(E) carbon tetrachloride.

ANSWER KEY

1.	(D)	6.	(B)
2.	(A)	7.	(B)
3.	(B)	8.	(E)
4.	(C)	9.	(E)
5.	(A)	10.	(C)

CHAPTER 3

Alkenes

Alkenes (olefins) are unsaturated hydrocarbons with one or more carbon-carbon double bonds. They have the general formula, C_nH_{2n}.

$$R_2C{=}CR_2 \qquad R = \text{Alkyl or H}$$

Example

$$H_2C{=}CH_2 \qquad\qquad H_2C{=}CH{-}CH_3$$

ethylene propylene

3.1 Nomenclature (IUPAC System)

A) Select the longest continuous chain of carbons containing the double bond. This is the parent structure and is assigned the name of the corresponding alkane with the suffix changed from "-ane" to "-ene."

B) Number the chain so that the position of the double bond is designated by the lowest possible number assigned to the first doubly bonded carbon.

Example

```
  5      4       3      2      1
CH3 - CH2 - CH = CH - CH3
         |
         Br
```

4-bromo – 2-pentene

Some common names given to families of alkenes are:

$H_2C = CH - R$	vinyl
$H_2C = CH - CH_2 - R$	allyl
$H_3C - CH = CH - R$	propenyl

Problem Solving Examples:

Write the IUPAC names for each of the following chemical structures:

```
(a) CH3-CH2-CH2-CH-CH-C=CH2
                |  |  |
                I  I  I

                      OH CH3
                       |  |
(b) CH3-CH2-CH2-CH=C-C----CH3
                          |
                          CH2-CH3

              CH2CH3
               |
(c) CH3-CH=C-CH2-CH-CH=CH2
           |          |
           CH2-CH3    Br

                      CH2-CH3  Cl
                       |        |
(d) CH3-CH2-CH=CH-C------CH=C-CH3
                       |
                       CH2
                       |
                       CH2
```

To write the IUPAC names for these alkenes, use the following rules:

(1) Select the parent structure of each compound by choosing the longest chain that contains the carbon-carbon double bond.

(2) Denote by number the position of the double bond in the parent chain. Designate its position by the number of the first doubly bonded carbon encountered when numbering from the end of the chain nearest the double bond.

(3) Indicate the positions of all functional groups attached to the parent chain by number.

With this in mind, the names of the compounds become:

(a) 2,3,4-triiodo-l-heptene

(b) 3,3-dimethyl-4-hydroxy-4-octene

(c) 3,5-diethyl-2-bromo-l,5-heptadiene

(d) 4,4-diethyl-2-chloro-2,5-octadiene

To understand the reasoning behind this naming, consider (b):

```
                        OH CH3
                        |  |
CH3-CH2-CH2-CH=C-C——CH3
                           |
                           CH2CH3
```

The longest continuous chain that contains the carbon-carbon double is eight carbons in length. Hence, the name must include the word "octene." The number of the first doubly bonded carbon encountered is four. (In this instance, numbering can be from either end because the double bond is located exactly equidistant from the ends.) Consequently, one has 4-octene. At this point, consider the substituents on the parent chain. Two methyl groups (CH_3) and a hydroxyl group (OH) are present. The methyls are said to be on the third carbon, while the hydroxyl is on the fourth carbon. In this fashion, the lowest possible numbers are used. (For example, if numbering started from the other end of the molecule, the hydroxyl would be on the fifth carbon and the methyls on the sixth carbon.) Overall, then, the name becomes 3,3-dimethyl-4-hydroxy-4-octene.

The other names of the compounds follow readily from this process.

Write the chemical structures for each compound listed.

(a) 1-hexene,
(b) 3-methyl-l-butene,
(c) 2,4-hexadiene,
(d) l-iodo-2-methyl-2-pentene,
(e) 2-chloro-3-methyl-2-hexene,
(f) 6,6-dibromo-5-methy1-5-ethyl-2,3-heptadiene

(1) Look at the complete name of the compound and pick out the parent name. It denotes the longest continuous chain that contains the carbon-carbon double bond. (All of the structures are alkenes.)

(2) Write out the carbon skeleton that makes up the parent chain. Determine the number of double bonds present by examining the suffix of the parent name. For example, "ene" means one double bond, where as "diene" means two.

(3) Position the double bond (or bonds) in the carbon skeleton as specified by the number directly (usually) in front of the parent name. For example, if the compound is 2-pentene, one would write

```
C – C – C = C – C
5   4   3   2   1
```

(Recall, the position of the double bond is given by the number of the first doubly bonded carbon encountered when numbering from the end of the chain nearest the double bond.)

(4) Position the functional group substituents on the chain as specified by the number directly in front. For example, 3-methyl-2-pentene would be:

```
        C
        |
C–C–C=C–C
5 4 3 2 1
```

The structures of the compounds in (a)–(f) become:

(a) $CH_3CH_2CH_2CH_2CH = CH_2$

(b)
```
CH2=CH-CH-CH3
       |
       CH3
```

(c) $CH_3 - CH = CH - CH = CH - CH_3$

(d)
```
CH3CH2CH=C—CH2I
         |
         CH3
```

(e)
```
CH3CH2CH2C==C-CH3
          |  |
        CH3  Cl
```

(f)
```
                 CH3  Br
                 |    |
CH3-CH=C=CH-C—————C-CH3
                 |    |
                 CH2  Br
                 |
                 CH3
```

To see how this process works, examine how structure (f) was written. The parent name is heptadiene. The prefix "hepta" indicates that seven carbons are present in the skeleton: C – C – C – C – C – C – C. The 2,3 indicates the positions of the two double bonds—it is a diene. Compounds with cumulated double bonds are called allenes. So, one can write

```
C-C=C=C-C-C-C
1 2 3 4 5 6 7
```

With this numbering system, the substituents are now added as specified by the 6 for the bromines and 5 for the methyl (CH_3) and ethyl groups.

```
        CH3  Br
        |    |
C-C=C=C-C————C-C
1 2 3 4 |    | 7
        CH2  Br
        |
        CH3
```

And now only the hydrogens need be added to obtain:

```
                 CH3  Br
                  |    |
CH3CH=C=CHC ——C-CH3
                  |    |
                 CH2  Br
                  |
                 CH3
```

3.2 Physical Properties of Alkenes

A) Alkenes have lower densities than that of water.

B) C_1 to C_4 are gases.

C) C_5 to C_{15} are liquids.

D) $> C_{16}$ are solids.

E) The boiling point, melting point, viscosity, and specific gravity increase with an increase in the length of the carbon chain.

F) Branching lowers the boiling point of alkenes.

G) Alkenes are relatively insoluble in water, but are soluble in nonpolar solvents such as benzene, ether, and chloroform.

H) Alkenes are colorless.

I) Alkenes show relatively higher reactivity than alkanes.

Problem Solving Example:

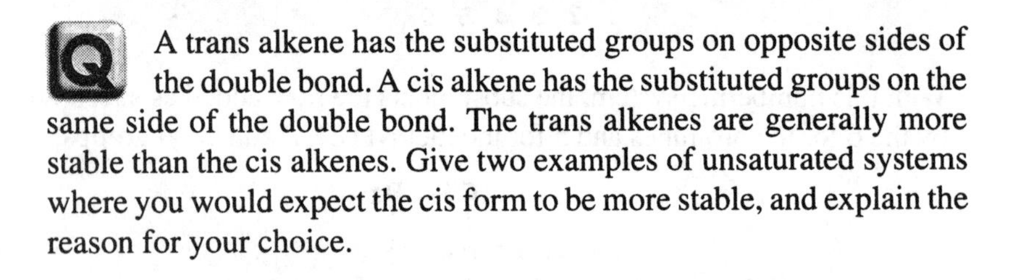

Q A trans alkene has the substituted groups on opposite sides of the double bond. A cis alkene has the substituted groups on the same side of the double bond. The trans alkenes are generally more stable than the cis alkenes. Give two examples of unsaturated systems where you would expect the cis form to be more stable, and explain the reason for your choice.

A To solve this problem, look for factors affecting stability of conformations. Two such factors include angle strain and van der Waals strain (steric strain).

Any atom tends to have bond angles that match those of its bonding orbitals: tetrahedral (109.5°) for sp^3 – hybridized carbon, for example. If there is a deviation from the "normal" bond angles, then angle strain results.

Nonbonded groups or atoms that just touch each other—that is, that are about as far apart as the sum of their van der Waals radii—attract each other. If the groups or atoms are brought any closer together, they repel each other. This crowding produces steric strain.

Consequently, when selecting two examples where a cis form will be more stable than a trans form, consider a situation where the trans conformation produces angle and/or steric strain and the cis does not, or does so to a lesser extent.

One example is cycloalkenes with small to medium-sized rings. Cyclobutene fits in this category. The cis and trans forms are illustrated here.

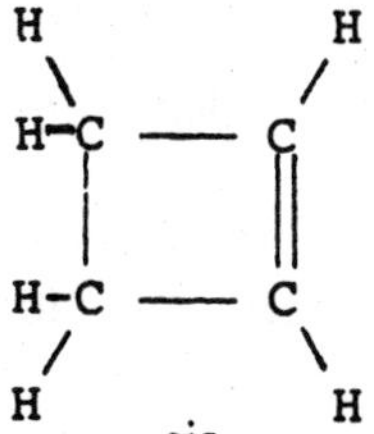

```
H           H                 H          H
 \         /                   \        /
H-C ——— C                       C ——— C          H
  |       ||                   /      ||        /
  |       ||                  H       C ——— C
H-C ——— C                            /        \
 /         \                        H          H
H           H                         trans
      cis
```

As can be seen in the diagram, the trans form would possess tremendous angle strain. Hence, the cis conformation would be more stable.

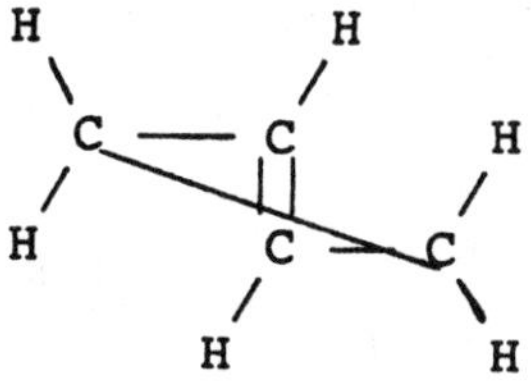

The second example of a situation where cis is favored over trans occurs in compounds such as:

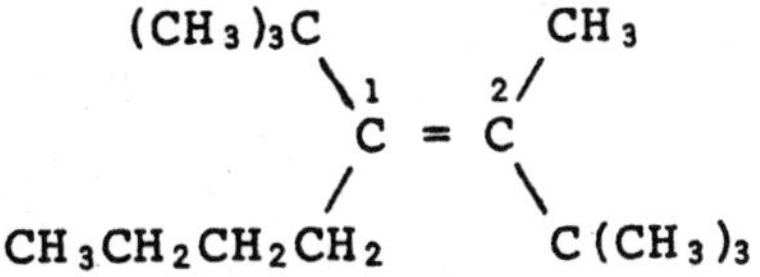

```
        (CH3)3C          CH3
              \1       2/
               C   =   C
              /         \
CH3CH2CH2CH2             C(CH3)3
```

Note that the highest priority group of the carbon labeled 1, $CH_3CH_2CH_2CH_2$, and the highest priority group of the carbon labeled 2, $C(CH_3)_3$, are on the same side of the molecule. Hence, this is the cis (NOT trans) conformation. This conformation is favored in stability over the trans form shown:

$$\begin{array}{ccc} CH_3CH_2CH_2CH_2 & & CH_3 \\ & \diagdown \quad \diagup & \\ & C = C & \\ & \diagup \quad \diagdown & \\ (CH_3)_3C & & C(CH_3)_3 \end{array}$$

The reason stems from the fact that the tertiary butyl group,

$$C(CH_3)_3 \left(H_3C-\overset{\displaystyle CH_3 \atop |}{C}-CH_3 \right)$$

is more bulky than $CH_3CH_2CH_2CH_2$ and, as such, creates more steric strain when positioned next to another $C(CH_3)_3$ as is the situation in the trans conformation.

3.3 Preparation of Alkenes

A) Dehydrohalogenation of Alkyl Halides

$$-\underset{H}{\underset{|}{\overset{|}{C}}}-\underset{X}{\underset{|}{\overset{|}{C}}}- + \underset{\text{strong base}}{KOH} \xrightarrow{\text{Alc.}} -\overset{|}{C}=\overset{|}{C}- + KX + H_2O$$

Example

$$\underset{\text{1-bromopropane}}{CH_3CH_2CH_2Br} \xrightarrow[\text{alc.}]{KOH} \underset{\text{propylene}}{CH_3CH=CH_2} + KBr + H_2O$$

Ease of dehydrohalogenation of alkyl halides is 3° > 2° > 1°.

B) Dehalogenation of Vicinal Dihalides

$$-\underset{X}{\underset{|}{\overset{|}{C}}}-\underset{X}{\underset{|}{\overset{|}{C}}}- + Zn \rightarrow -\overset{|}{C}=\overset{|}{C}- + ZnX_2$$

Example

$$\underset{\text{1,2-dibromopropane}}{CH_3\underset{Br}{\underset{|}{CH}}-\underset{Br}{\underset{|}{CH_2}}} + Zn \rightarrow \underset{\text{propylene}}{CH_3CH=CH_2} + ZnBr_2$$

C) Dehydration of Alcohols

$$\underset{\mathrm{H}\quad\mathrm{OH}}{-\overset{|}{\underset{|}{C}}-\overset{|}{\underset{|}{C}}-}\ \xrightarrow[\text{Heat}]{\text{Acid}}\ -\overset{|}{C}=\overset{|}{C}-\ +\ H_2O$$

Example

$$\underset{\text{ethanol}}{CH_3CH_2OH}\ \xrightarrow[170°C]{H_2SO_4}\ \underset{\text{ethylene}}{CH_2=CH_2}$$

Ease of dehydration of alcohols is 3° > 2° > 1°.

D) Pyrolysis (Cleavage by Heat)

$$\text{Alkane}\ \xrightarrow[\text{without catalyst}]{400\text{–}600°C;\ \text{with or}}\ \text{smaller alkane + alkene} + H_2$$

Example

$$\underset{\text{hexane}}{CH_3CH_2CH_2CH_2CH_2CH_3} \rightarrow \underset{\text{butane}}{CH_3CH_2CH_2CH_3} + \underset{\text{ethylene}}{CH_2 = CH_2} + H_2$$

E) Catalytic Hydrogenation and Reduction of Alkynes

$$R-C\equiv C-R\ \xrightarrow[\text{Pd or Ni-B(P-2)}]{H_2}\ \text{(R)(H)C=C(R)(H)}\quad \textit{cis}$$

$$R-C\equiv C-R\ \xrightarrow{\text{Na or Li, }NH_3}\ \text{(R)(H)C=C(H)(R)}\quad \textit{trans}$$

Problem Solving Examples:

How are alkenes prepared?

There are four basic methods of preparing alkenes, which are as follows:

(1) Dehydrohalogenation of alkyl halides

(2) Dehydration of alcohols

(3) Dehalogenation of vicinal dihalides

(4) Reduction of alkynes

The most important of these methods of preparation—since they are the most generally applicable—are the dehydrohalogenation of alkyl halides and the dehydration of alcohols.

Alkenes containing up to five carbon atoms can be obtained in pure form from the petroleum industry. Pure samples of more complicated alkenes must be prepared by methods like those outlined above.

The introduction of a carbon-carbon double bond (C = C) into a molecule containing only single bonds must necessarily involve the elimination of atoms or groups from two adjacent carbons:

$$-\underset{Y}{\overset{|}{\underset{|}{C}}}-\underset{Z}{\overset{|}{\underset{|}{C}}}- \longrightarrow -\overset{|}{C}=\overset{|}{C}- \quad \text{or} \quad -\underset{H}{\overset{|}{\underset{|}{C}}}-\underset{H}{\overset{|}{\underset{|}{C}}}- \xrightarrow[\text{Pt}]{\text{heat}} -\overset{|}{C}=\overset{|}{C}- + H_2$$

The elimination reactions not only can be used to make simple alkenes, but also provide the best general ways to introduce carbon-carbon double bonds into molecules of all kinds.

The four basic methods of preparing alkenes are outlined below:

(1) Dehydrohalogenation of alkyl halides

$$-\underset{H}{\overset{|}{\underset{|}{C}}}-\underset{X}{\overset{|}{\underset{|}{C}}}- + KOH \xrightarrow{\text{alc.}} -\overset{|}{C}=\overset{|}{C}- + KX + H_2O$$

The ease of dehydrohalogenation: 3 °> 2° > 1°

(2) Dehydration of alcohols

$$-\underset{H}{\overset{|}{\underset{|}{C}}}-\underset{OH}{\overset{|}{\underset{|}{C}}}- \xrightarrow{\text{acid}} -\overset{|}{C}=\overset{|}{C}- + H_2O$$

The ease of dehydration: 3° > 2° > 1°

(3) Dehalogenation of vicinal dihalides

$$-\underset{X}{\overset{|}{\underset{|}{C}}}-\underset{X}{\overset{|}{\underset{|}{C}}}- + Zn \longrightarrow -\overset{|}{C}=\overset{|}{C}- + ZnX_2$$

(4) Reduction of alkynes

$$R\text{-}C\equiv C\text{-}R \begin{cases} \xrightarrow[\text{Pd or Ni-B (P-2)}]{H_2} & \underset{H\quad H}{\overset{R\quad R}{C=C}} \quad \text{cis} \\ \xrightarrow{\text{Na or Li, } NH_3} & \underset{H\quad R}{\overset{R\quad H}{C=C}} \quad \text{trans} \end{cases}$$

Describe dehydrohalogenation of alkyl halides to obtain an alkene.

A Alkyl halides are converted into alkenes by the process known as dehydrohalogenation, which is the elimination of the halogen atom together with a hydrogen atom from a carbon adjacent to the one bearing the halogen. The general equation for this process is:

$$\underset{H\ \ X}{\text{-}\overset{|}{C}\text{-}\overset{|}{C}\text{-}} + \text{KOH (alcoholic)} \longrightarrow \text{-}\overset{|}{C}=\overset{|}{C}\text{-} + KX + H_2O$$

The alkene is prepared by heating together the alkyl halide and a solution of potassium hydroxide in alcohol. For example:

(1)

$$\underset{\text{n-Propyl chloride}}{CH_3CH_2CH_2Cl} \xrightarrow{\text{KOH (alc)}} \underset{\text{Propylene}}{CH_3CH=CH_2} + KCl + H_2O$$

(2)

$$\underset{\text{n-Butyl chloride}}{CH_3CH_2CH_2CH_2Cl} \xrightarrow{\text{KOH (alc)}} \underset{\text{1-Butene}}{CH_3CH_2CH=CH_2} + KCl + H_2O$$

In dehydrohalogenation, the more stable the alkene, the more easily it is formed as can be seen by the fact that the sequence for ease of formation of alkenes, $R_2C = CR_2 > R_2C = CHR > R_2C = CH_2$, $RCH = CHR > RCH = CH_2$, is the same for stability of alkenes. As one moves through the series of alkyl halides, that is, from 1° to 2° to 3°, the structure becomes more branched at the carbon carrying the halogen. Branching provides a greater number of hydrogens for attack by base, and hence a more favorable probability factor toward elimination. It leads to a more highly branched, more stable alkene, and hence a more stable transition state and lower E_{act}. The result is that in dehydrohalogenation the order of reactivity of RX is 3° > 2° > 1°.

The orientation of elimination reflects the stability of the alkene that would result. For example:

$$H_3CH_2CHBrCH_3 \xrightarrow{\text{KOH (alc)}} \underset{81\%}{CH_3CH{=}CHCH_3} \text{ and } \underset{19\%}{CH_3CH_2CH{=}CH_2}$$

2-butene (81%) predominates over 1-butene (19%) because the preferred alkene is the one with the greater number of alkyl groups attached to the doubly bonded carbon atoms as shown above in the sequence for ease of formation of alkenes.

3.4 Reactions of Alkenes

A) Hydroxylation (Glycol Formation)

$$\gt C{=}C\lt \xrightarrow[\text{or } HCO_2OH]{\text{cold alkaline } KMnO_4} -\underset{OH}{\underset{|}{C}}-\underset{OH}{\underset{|}{C}}-$$

B) Polymerization

Example

$$\underset{\text{Ethylene}}{n\,CH_2 = CH_2} \xrightarrow[\text{pressure}]{O_2,\ \text{Heat}} \underset{\text{Polyethylene}}{(-CH_2-CH_2-)_n}$$

C) Addition of Hydrogen Halides

$$-C{=}C- + HX \rightarrow -\underset{H}{\underset{|}{C}}-\underset{X}{\underset{|}{C}}- \qquad X = Cl,\ Br,\ I$$

In Markovnikov addition the hydrogen of the acid attaches itself to the carbon atom that has the greater number of hydrogens. In anti-Markovnikov addition the hydrogen of the acid attaches itself to the carbon atom that has the least number of hydrogens. Addition of hydrogen chloride or iodide follows Markovnikov's rule. Addition of hydrogen bromide will follow either rule, solely determined by the presence or absence of peroxides.

Example

$$CH_3CH=CH_2 \xrightarrow{HI} CH_3CHICH_3$$

propene 2-iodopropane

$$CH_3CH=CH_2 \xrightarrow{HBr} \begin{cases} \xrightarrow{\text{no peroxides}} CH_3CHBrCH_3 \text{ (2-bromopropane)} & \text{Markovnikov addition} \\ \xrightarrow{\text{peroxides}} CH_3CH_2CH_2Br \text{ (1-bromopropane)} & \text{Anti-Markovnikov addition} \end{cases}$$

D) Hydroboration-Oxidation

$$-\overset{|}{C}=\overset{|}{C}- + (BH_3)_2 \rightarrow -\underset{H}{\overset{|}{\underset{|}{C}}}-\underset{B}{\overset{|}{\underset{|}{C}}}- \xrightarrow[OH^-]{H_2O_2} -\underset{H}{\overset{|}{\underset{|}{C}}}-\underset{OH}{\overset{|}{\underset{|}{C}}}-$$

diborane

Anti-Markovnikov addition

E) Oxymercuration-Demercuration

$$-\overset{|}{C}=\overset{|}{C}- + H_2O + Hg(OAc)_2 \rightarrow -\underset{OH}{\overset{|}{\underset{|}{C}}}-\underset{HgOAc}{\overset{|}{\underset{|}{C}}}- \xrightarrow{NaBH_4} -\underset{OH}{\overset{|}{\underset{|}{C}}}-\underset{H}{\overset{|}{\underset{|}{C}}}-$$

Markovnikov addition

F) Addition of Halogens

$$-\overset{|}{C}=\overset{|}{C}- + X_2 \rightarrow -\underset{X}{\overset{|}{\underset{|}{C}}}-\underset{X}{\overset{|}{\underset{|}{C}}}- \qquad X = Cl_2, Br_2$$

Example

$$CH_3CH=CH_2 \xrightarrow{Br_2+CCl_4} CH_3CHBrCH_2Br$$

propene 1,2-dibromopropane

G) Catalytic Hydrogenation (Addition of Hydrogen)

$$-\overset{|}{C}=\overset{|}{C}- \; + \; H_2 \xrightarrow[\text{or Ni}]{Pt_1,\, Pd_1} -\underset{H}{\overset{|}{\underset{|}{C}}}-\underset{H}{\overset{|}{\underset{|}{C}}}-$$

Example

$$\underset{\text{propene}}{CH_3CH = CH_2} \xrightarrow{H_2,\,Ni} \underset{\text{propane}}{CH_3CH_2CH_3}$$

Problem Solving Examples:

What are addition reactions of alkenes?

A To understand addition reactions of alkenes, there must first be a good understanding of the nature of the double bond. The double bond consists of a strong sigma (σ) bond and a weak pi (π) bond. Therefore, it is expected that the reaction would involve breakage of the weaker bond. The typical reactions of the double bond are of the sort where the weaker π bond is broken and stronger σ bonds are formed:

$$-\overset{|}{C}=\overset{|}{C}- \quad + \quad YZ \longrightarrow -\underset{Y}{\overset{|}{\underset{|}{C}}}-\underset{Z}{\overset{|}{\underset{|}{C}}}- \qquad \text{Addition}$$

A reaction in which two molecules combine to yield a single molecule of product is called an addition reaction. The various types of addition reactions are:

(1) Catalytic Hydrogenation—Addition of Hydrogen

(2) Addition of Halogens

(3) Addition of Hydrogen Halides

(4) Addition of Sulfuric Acid

(5) Addition of Water—Hydration

(6) Halohydrin Formation

(7) Dimerization

(8) Alkylation

(9) Oxymercuration—Demercuration

(10) Hydroboration—Oxidation

(11) Addition of Free Radicals

(12) Polymerization

(13) Addition of Carbenes

(14) Hydroxylation—Glycol Formation

Predict the products of the following reactants after addition of hydrogen halides:

(a)
```
CH3C=CH-CH3
   |
   CH3
```
2-Methyl-2-Butene

(b) $CH_3CH{=}CHCH_3$

2-Butene

(c)
```
CH3CH=C — CH—CH3
      |    |
      CH3  CH3
```
3,4-Dimethyl-2-pentene

(d) $CH_3CH{=}CH_2$

Propylene

Alkenes can be converted to saturated compounds by addition of hydrogen halides such as hydrogen chloride, hydrogen bromide, or hydrogen iodide. The general reaction for this process is

```
 | |                  | |
-C=C-  + HX ———→     -C-C-    (HX = HCl, HBr, or HI)
                      | |
                      H X
```

The product is generally referred to as an alkyl halide, and it is produced by passing dry gaseous hydrogen halide through the alkene.

In ionic addition of an acid (such as hydrogen halide) to the carbon-carbon double bond of an alkene, the hydrogen of the acid attaches itself to the carbon atom that already holds the greater number of hydrogens. This principle is known as Markovnikov's rule. Remembering this principle, the products are written as:

(a)

$$CH_3-\underset{\underset{CH_3}{|}}{C}=CH-CH_3 + HI \longrightarrow CH_3-\overset{\overset{I}{|}}{\underset{\underset{CH_3}{|}}{C}}-CH_2-CH_3$$

2-Methyl-2-butene — tert-Pentyl iodide (2-Iodo-2-methylbutane)

(b)

$$CH_3CH=CHCH_3 + HBr \longrightarrow CH_3\underset{\underset{Br}{|}}{C}HCH_2CH_3$$

2-Butene — 2-Bromobutane

(c)

$$CH_3CH=\underset{\underset{CH_3}{|}}{C}-\underset{\underset{CH_3}{|}}{C}HCH_3 + HCl \longrightarrow CH_3-CH_2-\overset{\overset{Cl}{|}}{\underset{\underset{CH_3}{|}}{C}}-\underset{\underset{CH_3}{|}}{CH}-CH_3$$

3,4-Dimethyl-2-pentene — 3-Chloro-2,3-dimethylpentane

(d)

$$CH_3-CH=CH_2 + HI \longrightarrow CH_3-\overset{\overset{I}{|}}{C}H-CH_3$$

Propylene — Isopropyl iodide

3.5 Dienes

Dienes have the structural formula, C_nH_{2n-2}. In the IUPAC nomenclature system, dienes are named in the same manner as alkenes, except that the suffix "-ene" is replaced by "-diene," and two numbers must be used to indicate the position of the double bonds.

3.6 Classification of Dienes

$$-\overset{|}{C}=C=\overset{|}{C}-$$ Cumulated double bonds (allenes)

$$-\overset{|}{C}=\overset{|}{C}-\underset{|}{C}=\overset{|}{C}-$$ Conjugated (alternating) double bonds

$-\overset{|}{C}=\underset{|}{\overset{|}{C}}-(CH_2)_n-\overset{|}{C}=\overset{|}{C}-$ Isolated (nonconjugated) double bonds

Problem Solving Examples:

What are dienes? Briefly discuss their properties and system of nomenclature.

Dienes are alkenes that contain two carbon-carbon double bonds. These double bonds can be either adjacent (called cumulated double bonds), or separated by one carbon-carbon single bond (conjugated), or separated by two or more carbon single bonds (isolated).

The nomenclature of dienes is analogous to that of alkenes, with certain modifications to indicate the extra double bond. The numbering of carbons is the same as that of alkenes. The numbers of the olefinic carbons are placed at the beginning of the name, which is suffixed by -diene instead of -ene. For example, a compound of the structure

$$\overset{5}{C}H_3-\overset{4}{C}H=\overset{3}{C}H-\overset{2}{C}H=\overset{1}{C}H_2$$

is called 1,3-pentadiene, where "1,3" indicates the positions of the double bonds.

The properties of dienes depend upon the arrangement of the double bonds and the other substituents. If the double bonds are widely separated (as in isolated dienes), the compound behaves like a simple alkene. If the compound is conjugated, there is potential delocalization of the π electrons. Conjugated double bonds are, in general, more stable than isolated double bonds because of the resonance effect. Cumulated double bonds (allenes) are more unstable than isolated double bonds because of their geometry and electron density.

Of all the dienes studied so far, one of the most important is 1,3-butadiene, $CH_2 = CH - CH = CH_2$. It is used in making synthetic rubber by free-radical polymerization.

Name or draw structures for the following compounds.

(a) CH_3 (on a cyclopentadiene ring)

(b) $CH_3CH{=}CH{-}CH{=}CH{-}C(=O)OH$

(c) (cyclopentadiene ring)$=\overset{+}{N}=\ddot{N}^-$

(d) **2,3-dimethyl-1,4-hexadiene**

(e) **3,6-dinitro-1,4-cyclohexadiene**

A The IUPAC system of naming for dienes is similar to that for simple alkenes, except that it is prefixed by two numbers to indicate the positions of the double bonds, and suffixed by -diene instead of -ene. The numbers are assigned so as to have the lowest combination.

(a) (ring numbered 1–5, with CH_3 on C5) is a cyclic compound containing five carbons in the ring, a methyl substituent, and two carbon-carbon double bonds. The name of the compound is 5-methyl-1,3-cyclopentadiene.

(b) $\underset{6}{CH_3}{-}\underset{5}{CH}{=}\underset{4}{CH}{-}\underset{3}{CH}{=}\underset{2}{CH}{-}\underset{1}{C}(=O)OH$

contains six carbons in a chain, two carbon-carbon double bonds, and a carboxyl functional group. Since the COOH group takes precedence over diene in naming, the compound is named as a derivative of hexanoic acid. The numbering of carbons in carboxylic acid starts with the carbonyl carbon as number one. Thus, the name of the compound is 2,4-dienehexanoic acid.

(c) 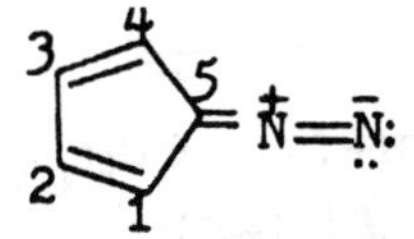 [structure: cyclopentadiene ring numbered 1–5, with $\overset{+}{N}=\overset{-}{N}:$ at carbon 5]

is a cyclic diene with the double bonds at carbons 1 and 3. At carbon number 5 there is a diazo substituent. The IUPAC name for the compound is 5-diazo-1, 3-cyclopentadiene.

(d) 2,3-dimethyl-1,4-hexadiene has a straight chain of six carbons with double bonds at carbons 1 and 4. At carbons 2 and 3, there are methyl substitutions. Therefore, the structure of the compound is:

$$CH_3-CH=\overset{4}{C}H-\underset{CH_3}{\overset{3}{C}H}-\underset{CH_3}{\overset{2}{C}H}=\overset{1}{C}H$$

(e) 3,6-dinitro-1,4-cyclohexadiene is a six-carbon cyclic compound with double bonds at carbon numbers 1 and 4. At carbons 3 and 6, there are two nitro substitutions. The structure of the compound is:

[structure: six-membered ring numbered 1–6, with double bonds at 1–2 and 4–5, and NO_2 at carbons 3 and 6]

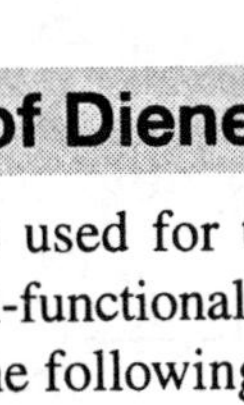

3.7 Preparation of Dienes

All preparation methods used for the alkenes may be used for nonconjugated dienes using di-functional starting materials. Conjugated dienes may be produced in the following ways:

A) Dehydration of 1,3-Diols

$$CH_3-\underset{OH}{CH}-CH_2-\underset{OH}{CH_2} \xrightarrow[\text{Acid}]{\text{Heat}} CH_2=CH-CH=CH_2 + 2H_2O$$

1,3-butanediol → 1,3-butadiene

B) Dehydrogenation

$$CH_3-CH_2-CH_2-CH_3 \xrightarrow[\text{catalyst}]{\text{Heat}} \begin{cases} CH_3-CH_2-CH=CH_2 \\ CH_2=CH-CH=CH_2 \\ CH_3-CH=CH-CH_3 \end{cases}$$

(The 1-butene and 2-butene products are converted further to $CH_2=CH-CH=CH_2$ by Heat/Catalyst.)

Allene may be produced from glycerol by stepwise substitution and eliminations.

$$\begin{matrix} CH_2-OH \\ | \\ CH-OH \\ | \\ CH_2-OH \end{matrix} \xrightarrow{HBr} \begin{matrix} CH_2-Br \\ | \\ CH-Br \\ | \\ CH_2-Br \end{matrix} \xrightarrow[\text{Alc.}]{KOH} \begin{matrix} CH_2 \\ \| \\ C-Br \\ | \\ CH_2-Br \end{matrix} \xrightarrow[\text{alc.}]{Zn} \begin{matrix} CH_2 \\ \| \\ C \\ \| \\ CH_2 \end{matrix}$$

Problem Solving Example:

Predict the major products of dehydrohalogenation of the following compounds.

(a) $C_6H_5-CH_2-\underset{Br}{\underset{|}{CH}}-CH_3$

(b) $CH_3-\underset{Br}{\underset{|}{CH}}-CH_2-CH=CH_2$

(c) $CH_3-\overset{CH_3}{\overset{|}{\underset{Br}{\underset{|}{C}}}}-CH_2-CH=CH_2$

The more stable conjugated diene will be produced upon dehydrohalogenation.

(a) $C_6H_5-CH_2-\underset{Br}{\underset{|}{CH}}-CH_3 \longrightarrow C_6H_5-CH=\underset{H}{C}-CH_3$

(b) $CH_3-\underset{Br}{\underset{|}{CH}}-CH_2-CH=CH_2 \longrightarrow CH_3-CH=CH-CH=CH_2$

(c) $CH_3-\overset{CH_3}{\overset{|}{\underset{Br}{\underset{|}{C}}}}-CH_2-CH=CH_2 \longrightarrow CH_3-\overset{CH_3}{\overset{|}{C}}=CH-CH=CH_2$

3.8 Reactions of Dienes

A) 1,4-additions to give 1,4-alkadienes

Example

$$CH_2 = CH{-}CH = CH_2 + Br_2 \rightarrow \underset{\displaystyle Br}{\underset{|}{CH_2}}{-}CH = CH{-}\underset{\displaystyle Br}{\underset{|}{CH_2}}$$

1,3-butadine 1,4-dibromobutene

B) Isomerization of certain dienes to give alkynes

Example

$$(CH_3)_2C = C = CH_2 + Na \rightarrow (CH_3)_2CH - C \equiv CH$$

C) Polymerization to give synthetic rubber

Example

$$n\,CH_2 = \overset{\displaystyle CH_3}{\overset{|}{C}}{-}CH = CH_2 \xrightarrow{\text{Catalyst}} \text{Rubber-like products}$$

isoprene

$$n\,CH_2 = \overset{\displaystyle Cl}{\overset{|}{C}}{-}CH = CH_2 \longrightarrow \text{Neoprene}$$

chloroprene

Problem Solving Examples:

Q Account for the fact that 2-methy1-1,3-butadiene reacts (a) with HCl to yield only 3-chloro-3-methyl-l-butene and l-chloro-3-methyl-2-butene; and (b) with bromine to yield only 3,4-dibromo-3-methyl-l-butene and 1,4-dibromo-2-methyl-2-butene.

(a) In acid solution, one of the double bonds in a diene becomes protonated. As in the case of alkenes, protonation will occur to form the most stable cation. 2-methy1-1,3-butadiene is protonated at carbon 1 because it forms the most stable cation; there is electron delocalization involving the remaining double bond that stabilizes the cation:

$$CH_2=\underset{\substack{|\\CH_3}}{C}-CH=CH_2 \xrightarrow{H^+} CH_3-\overset{CH_3}{\underset{+}{C}}-CH=CH_2$$

$$CH_3-\overset{CH_3}{\underset{+}{C}}-CH=CH_2 \longleftrightarrow CH_3-\overset{CH_3}{C}=CH-\underset{+}{CH_2}$$

Had protonation occurred at carbons 2 or 3, no electron delocalization would have been possible; the cation would be relatively unstable. Had protonation occurred at carbon 4, electron delocalization is possible. However, protonation of carbon 4 distributes the positive charge between a secondary and primary carbon, whereas protonation of carbon 1 distributes the charge between a tertiary and primary carbon.

$$CH_2=\overset{CH_3}{C}-CH=CH_2 \xrightarrow{H^+} CH_2=\overset{CH_3}{C}-\underset{+}{CH}-CH_3$$

$$CH_2=\overset{CH_3}{C}-\overset{+}{CH}-CH_3 \longleftrightarrow \overset{+}{CH_2}-\overset{CH_3}{C}=CH-CH_3$$

Above: Protonation and stabilization at carbon 4.

Since tertiary carbon stabilizes a positive charge more than a secondary carbon, the cation formed by protonation of carbon 1 is more stable than the one formed by protonation of carbon 4. Hence, 2-methyl-1,3-butadiene will be protonated by HC1 as follows:

$$Cl-H + CH_2=\overset{CH_3}{C}-CH=CH_2 \rightarrow CH_3-\underset{\delta+}{\overset{CH_3}{C}}\cdots CH\cdots\underset{\delta+}{CH_2} + Cl^-$$

The chloride will react with the diene cation to form two products. This is because the positive charge is distributed between two carbons (2 and 4); hence, there are two sites to which the chloride could attach.

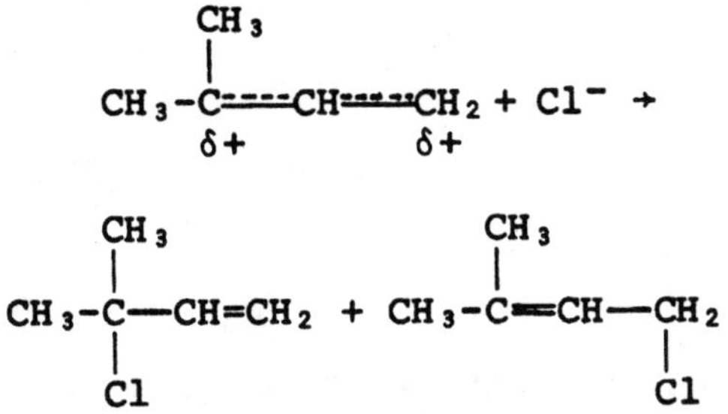

$$CH_3-\overset{\overset{CH_3}{|}}{\underset{\delta+}{C}}\text{=-=-}CH\text{=-=-}\underset{\delta+}{CH_2} + Cl^- \rightarrow$$

$$CH_3-\overset{\overset{CH_3}{|}}{\underset{\underset{Cl}{|}}{C}}-CH=CH_2 + CH_3-\overset{\overset{CH_3}{|}}{C}=CH-\underset{\underset{Cl}{|}}{CH_2}$$

The overall reaction can be written as:

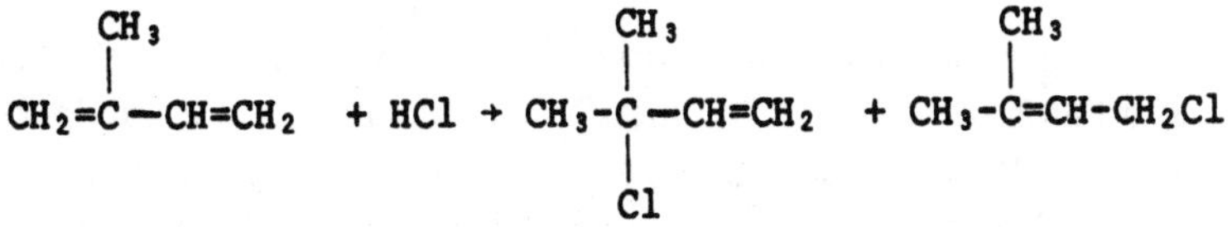

$$CH_2=\overset{\overset{CH_3}{|}}{C}-CH=CH_2 + HCl \rightarrow CH_3-\overset{\overset{CH_3}{|}}{\underset{\underset{Cl}{|}}{C}}-CH=CH_2 + CH_3-\overset{\overset{CH_3}{|}}{C}=CH-CH_2Cl$$

2-methyl-1,3-butadiene — 3-chloro-3-methyl-1-butene — 1-chloro-3-methyl-2-butene

3-chloro-3-methyl-l-butene is referred to as the "1,2" adduct. This is because addition of hydrogen halide took place at carbons 1 and 2. l-chloro-3-methyl-2-butene is referred to as the "1,4" adduct because addition occurred at carbons 1 and 4. The product distribution kinetically will favor the "1,2" adduct over the "1,4" adduct because the former actually resulted from a more stable carbonium ion than the latter, as shown by the following:

"1,2" adduct

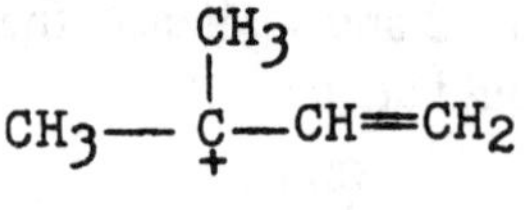

$$CH_3-\underset{+}{\overset{CH_3}{\overset{|}{C}}}-CH{=}CH_2 \quad 3° \text{ cation}$$

"1,4" adduct

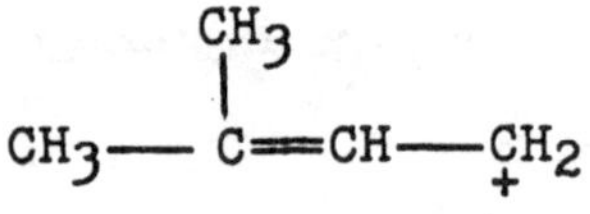

$$CH_3-\overset{CH_3}{\overset{|}{C}}{=}CH-\underset{+}{C}H_2 \quad 1° \text{ cation}$$

Products:

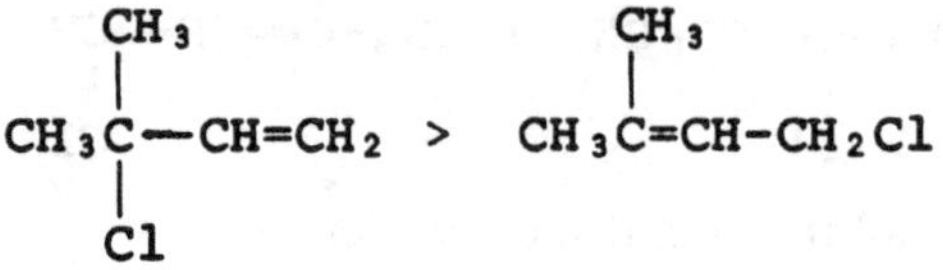

$$CH_3\underset{Cl}{\underset{|}{\overset{CH_3}{\overset{|}{C}}}}-CH{=}CH_2 > CH_3\overset{CH_3}{\overset{|}{C}}{=}CH-CH_2Cl$$

The product distribution will thermodynamically favor the "1,4" adduct over the "1,2" adduct. This is because an internal double bond ("1,4" adduct) is more stable than a terminal double bond ("1,2" adduct); that is, the more substituted an alkene, the more stable the molecule. Hence, the product distribution depends totally upon the reaction conditions.

(b) Reaction of 2-methyl-1,3-butadiene with bromine is similar to part (a) in that addition occurs at carbons 1 and 2 and at carbons 1 and 4. The difference is that in part (a) the diene undergoes a hydrohalogenation, whereas in part (b) the diene is halogenated. The electron-rich carbon 1-carbon 2 double bond polarizes the bromine molecule so that a Br_+ is added to carbon 1. The Br^- can then add to carbon 2 or carbon 4. The reaction occurs as:

$$CH_2{=}\overset{\displaystyle CH_3}{\overset{|}{C}}{-}CH{=}CH_2 + Br_2 \rightarrow \left[\underset{\displaystyle \overset{\delta+}{Br}{---}\overset{\delta-}{Br}}{CH_2}{\cdots}\overset{\displaystyle CH_3}{\overset{|}{C}}{-}CH{=}CH_2 \right] \longrightarrow BrCH_2{-}\underset{+}{\overset{\displaystyle CH_3}{\overset{|}{C}}}{-}CH{=}CH_2 + Br^-$$

$$BrCH_2{-}\underset{\delta+}{\overset{\displaystyle CH_3}{\overset{|}{C}}}{\cdots}CH{\cdots}\underset{\delta+}{CH_2} + Br^- \rightarrow BrCH_2{-}\underset{\displaystyle Br}{\underset{|}{\overset{\displaystyle CH_3}{\overset{|}{C}}}}{-}CH{=}CH_2 +$$

3,4-dibromo-3-methyl-1-butene

$$BrCH_2{-}\overset{\displaystyle CH_3}{\overset{|}{C}}{=}CH{-}CH_2Br$$

1,4-dibromo-2-methyl-2-butene

The overall reaction can be written as:

$$CH_2{=}\overset{\displaystyle CH_3}{\overset{|}{C}}{-}CH{=}CH_2 + Br_2 \rightarrow BrCH_2{-}\underset{\displaystyle Br}{\underset{|}{\overset{\displaystyle CH_3}{\overset{|}{C}}}}{-}CH{=}CH_2 + BrCH_2{-}\overset{\displaystyle CH_3}{\overset{|}{C}}{=}CH{-}CH_2Br$$

Q Formulate chain initiation, propagation, and termination steps for the polymerization of butadiene by a peroxide catalyst. Consider carefully possible structures for the growing-chain radical. Show the expected structure of the polymer.

A Peroxides are very unstable compounds that are often used as a catalyst for free radical polymerization reactions. The peroxide breaks down into two alkoxy radicals as shown:

$$ROOR' \longrightarrow RO\cdot + R'O\cdot$$

peroxide alkoxy radicals

This is the initiation step of a free radical polymerization process. Free radical polymerizations, like all addition polymerization reactions, produce head-to-tail polymerization; that is, the growing end of the polymer is the most stable possible radical. In the case of 1,3-butadiene the alkoxy radical attacks an end carbon and not an internal carbon. This is because attack of an end carbon forms a resonance stabilized free radical, whereas attack of an internal carbon forms a radical with no such stabilization. Hence, the end carbon will be attacked by the alkoxy radical to form the more stable species. This is shown as:

$$RO\cdot + CH_2{=}CHCH{=}CH_2 \rightarrow ROCH_2\text{-}\dot{C}H\text{-}CH{=}CH_2 \leftrightarrow ROCH_2\text{-}CH{=}CH\text{-}\dot{C}H_2$$

The resulting radical will attack an end carbon of another 1,3-butadiene molecule. There are three possible products, which all depend on where the end carbon of the diene is bonded to the radical. Note that the free electron of the radical $ROCH_2C\dot{H}CH = CH_2$ distributes its time between carbons 2 and 4 (where carbon 1 is the $RO - CH_2-$ carbon).

Polymerization can occur as a 1,2-addition or a 1,4-addition; that is, the radical can attack a diene molecule with the free electron on either carbon 2 or carbon 4. This accounts for two possible polymerization products, but there is a third one. Note that the growing polymer in 1,4-addition can have two different configurations: cis and trans. The radical can add to a diene by 1,4-addition to produce a trans or a cis product:

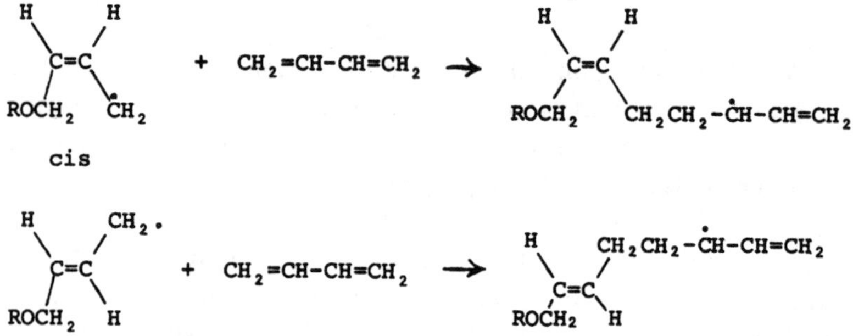

$$\underset{\text{cis}}{(H)(ROCH_2)C{=}C(H)(\dot{C}H_2)} + CH_2{=}CH\text{-}CH{=}CH_2 \rightarrow (H)(ROCH_2)C{=}C(H)(CH_2CH_2\text{-}\dot{C}H\text{-}CH{=}CH_2)$$

$$\underset{\text{trans}}{(H)(ROCH_2)C{=}C(CH_2\cdot)(H)} + CH_2{=}CH\text{-}CH{=}CH_2 \rightarrow (H)(ROCH_2)C{=}C(CH_2CH_2\text{-}\dot{C}H\text{-}CH{=}CH_2)(H)$$

1,4-addition

The radical adds to the diene by 1,2-addition as shown:

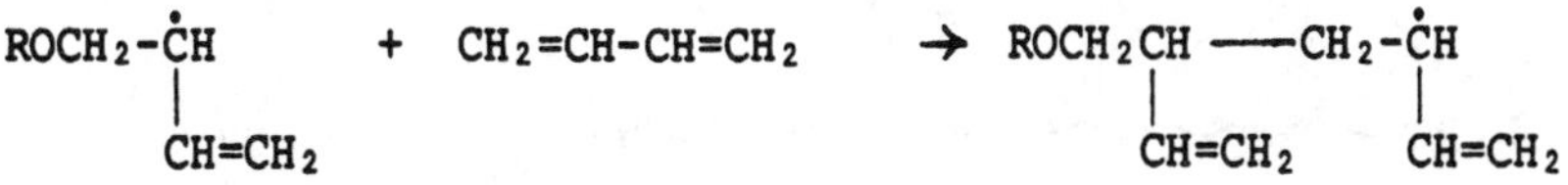

$$ROCH_2\text{-}\dot{C}H(CH{=}CH_2) \quad + \quad CH_2{=}CH{-}CH{=}CH_2 \quad \rightarrow \quad ROCH_2CH(CH{=}CH_2){-}CH_2{-}\dot{C}H(CH{=}CH_2)$$

The chain lengthening of the polymeric radical is the propagation step. The polymer can be formed as any combination of 1,2-, 1,4-cis, or 1,4-trans additions; the polymer can be a result of one or all of the three addition processes.

The termination step of a polymerization reaction puts a stop to the growing polymer. In free radical polymerization, the termination step rids the growing polymer of its free electron. This generally proceeds by any one of three different methods: dimerization, disproportionation, and abstraction. Dimerization involves the joining of two growing polymer radicals. It can be shown as:

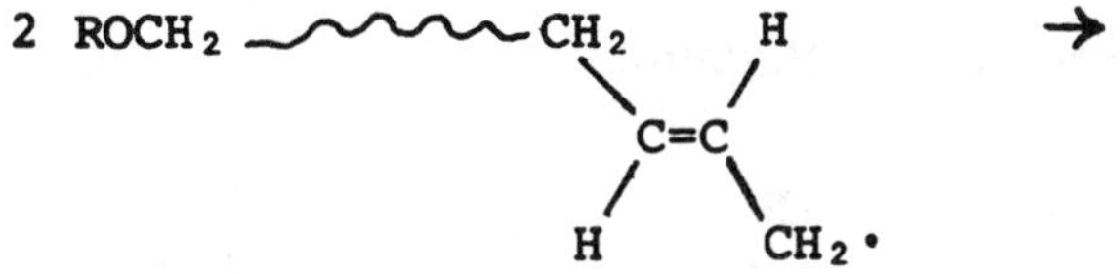

$$2\ ROCH_2{\sim\sim\sim}CH_2(H)C{=}C(H)CH_2\cdot \quad \rightarrow$$

growing polymer radical

$$ROCH_2{\sim\sim\sim}CH_2(H)C{=}C(H)CH_2{-}CH_2(H)C{=}C(H)CH_2{\sim\sim\sim}CH_2OR$$

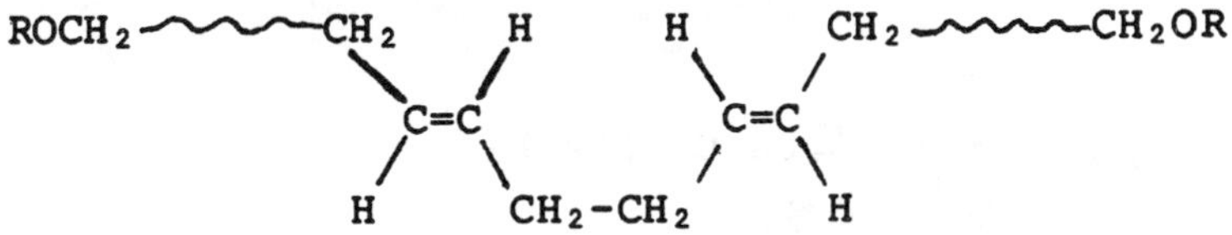

polymer (formed by dimerization)

Disproportionation reactions involve the transfer of a hydrogen radical from one growing polymer radical to another. This results in the joining of a hydrogen radical with a growing polymer radical and the formation of a carbon-carbon bond in the other growing polymer radical. This is shown as:

$$RO\sim\sim\sim CH_2-CH-CH=CH_2$$

$$RO\sim\sim\sim \ddot{C}H-\dot{C}H-CH=CH_2$$

$$\downarrow$$

$$RO\sim\sim\sim CH_2CH_2CH=CH_2 \quad +$$

$$RO\sim\sim\sim \dot{C}H-CH-CH=CH_2$$

$$\downarrow$$

$$RO\sim\sim\sim CH_2CH_2CH=CH_2 \quad +$$

$$RO\sim\sim\sim CH=CH-CH=CH_2$$

Abstraction involves the loss of a hydrogen atom to form a carbon-carbon bond. This is shown as:

$$RO\sim\sim\sim \ddot{C}H-CH-CH=CH_2$$

$$H \downarrow -H^{\cdot}$$

$$RO\sim\sim\sim \dot{C}H-CH-CH=CH_2$$

$$\downarrow$$

$$RO\sim\sim\sim CH=CH-CH=CH_2$$

CHAPTER 4

Alkynes

Alkynes are unsaturated hydrocarbons containing triple bonds. They have the general formula, C_nH_{2n-2}.

$R-C \equiv C-B$ $\qquad$ R = Alkyl or H

Example

$H-C \equiv C-H$ $\qquad$ Simplest alkyne

acetylene

4.1 Nomenclature (IUPAC System)

Alkynes are named in the same manner as alkenes, except that the suffix "-ene" is replaced with "-yne."

When both a double bond and a triple bond are present, the hydrocarbon is called an alkenyne. In this case, the double bond is given preference over the triple bond when numbering.

Example $\quad CH_3-C \equiv CH$ $\qquad CH_3-C \equiv C-CH = CH_2$

propyne $\qquad$ 1-penten-3-yne

Problem Solving Example:

What is an alkyne?

An alkyne is an organic compound distinguished by the carbon-carbon triple bond. The simplest member of the alkyne family is acetylene, with the formula C_2H_2 or $H - C \equiv C - H$. The general formula for alkynes is C_NH_{2N-2}.

The alkynes are named according to two systems:

(1) They are considered to be derived from acetylene by replacement of one or both hydrogen atoms by alkyl groups. For example,

(a) $HC \equiv CCH_2CH_3$ (b) $CH_3C \equiv CCH_3$

ethylacetylene dimethylacetylene

(2) For more complicated alkynes the IUPAC names are used. The rules are exactly the same as for the naming of alkenes, except that the ending -yne replaces -ene. The parent structure is the longest continuous chain that contains the triple bond, and the positions both of substituents and of the triple bond are indicated by numbers. The triple bond is given the number of the first triply bonded carbon encountered, starting from the end of the chain nearest the triple bond. For example,

```
CH3-C≡C-CH-CH3
        |
        CH3
```

4-methyl-2-pentyne

4.2 Physical Properties of Alkynes

A) Lower carbon members are gases with boiling points somewhat higher than corresponding alkenes.

B) Terminal alkynes have lower boiling points than isomeric internal alkynes.

C) The hydrogens in terminal alkynes are relatively acidic.

D) The dipole moment is small, but larger than that of an alkene.

E) Other physical properties are essentially the same as those for alkanes and alkenes.

Problem Solving Example:

How many sites of unsaturation does a compound with the formula $C_8H_{12}O_2$ contain?

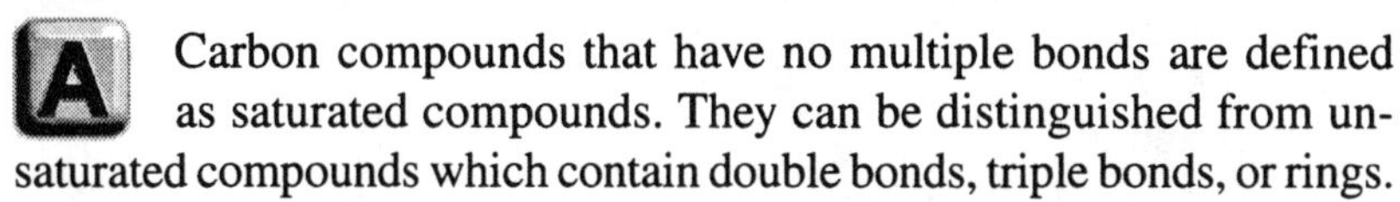

Carbon compounds that have no multiple bonds are defined as saturated compounds. They can be distinguished from unsaturated compounds which contain double bonds, triple bonds, or rings.

If it is first assumed that the compound is saturated, then we can predict the number of hydrogen atoms that must be present to meet this assumption. This is calculated by using the general formula of a saturated compound, C_nH_{2n+2}. Since the compound in question possesses eight carbon atoms, we expect 2(8) + 2 or 18 hydrogen atoms to be present if the compound is saturated. But in $C_8H_{12}O_2$ only 12 such atoms are present. The deficiency in hydrogen atoms is 18 – 12 = 6.

The degree of unsaturation is the number of double bonds, triple bonds, or rings present in a particular compound. For example, C_2H_6 (ethane), written as

```
  H H
  | |
H-C-C-H,
  | |
  H H
```

is saturated, where C_2H_4 (ethene), structural formula

```
H —C ═ C - H
   |   |
   H   H
```

has one double bond (one site of unsaturation) and two hydrogen atoms less than its fully saturated counterpart. So, two hydrogen atoms are associated with every site of unsaturation. Since each site of unsaturation is associated with the loss of two hydrogen atoms, the number of sites of unsaturation for $C_8H_{12}O_2$ must be 6 ÷ 2 or 3.

4.3 Preparation of Alkynes

A) Dehydrohalogenation of Alkyl Dihalides

$$-\underset{X}{\overset{H}{C}}-\underset{X}{\overset{H}{C}}- \xrightarrow[\text{Alc.}]{\text{KOH}} -\underset{X}{C}=\overset{H}{C}- \xrightarrow{NaNH_2} -C\equiv C-$$

Example

$$CH_3-\underset{Br}{CH}-\underset{Br}{CH_2} \xrightarrow[\text{Alc.}]{\text{KOH}} CH_3-CH=\underset{Br}{CH} \xrightarrow{NaNH_2} CH_3C\equiv CH$$

1,2-dibromopropane propyne

B) Dehalogenation of Tetrahalides

$$-\underset{X}{\overset{X}{C}}-\underset{X}{\overset{X}{C}}- + 2Zn \rightarrow -C\equiv C- + 2ZnX_2$$

Example

$$CH_3-\underset{Br}{\overset{Br}{C}}-\underset{Br}{\overset{Br}{CH}} \xrightarrow{2Zn} CH_3-C\equiv CH + 2ZnBr_2$$

1,1,2,2-tetrabromopropane propyne

C) Reaction of Water with Calcium Carbide

$$CaC_2 + H_2O \rightarrow CH\equiv CH + Ca(OH)_2$$

acetylene

Problem Solving Examples:

Prepare an alkyne from each alkene precursor by the dehydrohalogenation of alkyl dihalides method.

(a) $CH_3CH{=}CH_2$

propene

(b)
```
         CH3
         |
CH3CH2CHCH=CH2
```

3-methyl-1-pentene

A A carbon-carbon triple bond is formed in the same way as a double bond: elimination of atoms or groups from two adjacent carbons. The groups eliminated and the reagents used are essentially the same as in the preparation of alkenes.

In the first reaction of the synthesis, either Br_2 or Cl_{12} is bubbled through the precursor alkene yielding an alkyl dihalide,

```
 H H
 | |
-C-C- .
 | |
 X X
```

This product is then treated with an alcohol-potassium hydroxide solution to form a vinyl halide,

```
   H
   |
-C=C-
 |
 X
```

Sodamide ($NaNH_2$) is added to the vinyl halide to produce a carbon-carbon triple bond, an alkyne. Overall, the synthetic preparation is:

```
 H H           H H                   H
 | |           | |                   |
-C=C-  --X2-->  -C-C-  --KOH (alc)-->  -C=C-  --NaNH2-->  -C≡C-
               | |                 |
               X X                 X
```

Vinyl halides are very unreactive so that under mild conditions the dehydrohalogenation stops here. Only under more vigorous conditions —use of a strong base ($NaNH_2$)—is the alkyne generated.

Employing the general reaction of synthesis, problems (a) and (b) can be solved.

(a)

$$CH_3CH{=}CH_2 \xrightarrow{Br_2} CH_3\underset{Br}{CH}{-}\underset{Br}{CH_2} \xrightarrow{KOH\ (alc)} CH_3CH{=}CHBr \xrightarrow{NaNH_2} CH_3C{\equiv}CH$$

propene → 1,2-dibromo-propane → 1-bromo-1-propene → propyne

(b)

$$CH_3CH_2\overset{CH_3}{CH}CH{=}CH_2 \xrightarrow{Br_2} CH_3CH_2\overset{CH_3}{CH}\underset{Br}{CH}{-}\underset{Br}{CH_2} \xrightarrow{t\text{-}BuOH\ (alc)} CH_3CH_2\overset{CH_3}{CH}{-}\underset{Br}{C}{=}CH_2 \xrightarrow{NaNH_2} CH_3CH_2\overset{CH_3}{CH}C{\equiv}CH$$

3-methyl-pentene → 3-methyl-1,2-dibromo pentane → 3-methyl-2-bromo-1-pentene → 3-methyl-1-pentyne

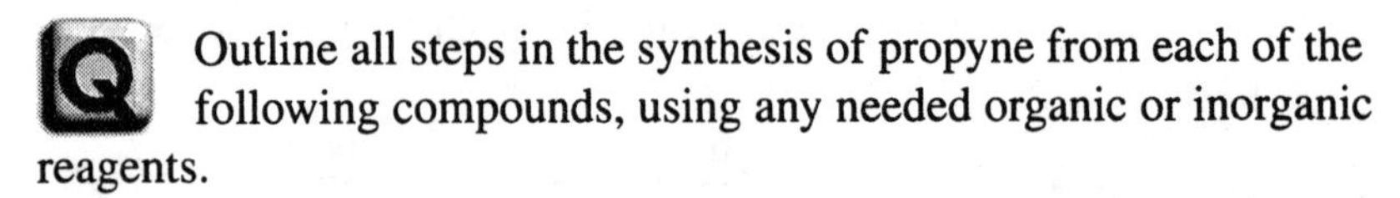

Q Outline all steps in the synthesis of propyne from each of the following compounds, using any needed organic or inorganic reagents.

(a) 1,2-dibromopropane	(e) n-propyl alcohol
(b) propylene	(f) 1,1-dichloropropane
(c) isopropyl bromide	(g) acetylene
(d) propane	(h) 1,1,2,2-tetrabromopropane

(a) 1, 2-dibromopropane:

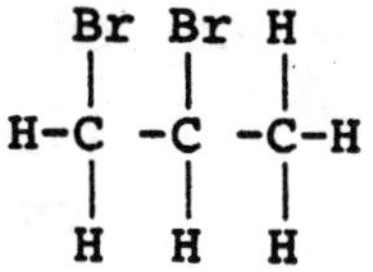

```
   Br Br H
   |  |  |
H-C -C -C-H      C3H6Br2
   |  |  |
   H  H  H
```

Propyne can be prepared by introducing a triple bond by eliminating two molecules of hydrogen bromide. The dehydrohalogenation proceeds in two steps. The first step involves the addition of alcoholic KOH. Sodium amide ($NaNH_2$) in liquid ammonia is used to create the triple bond:

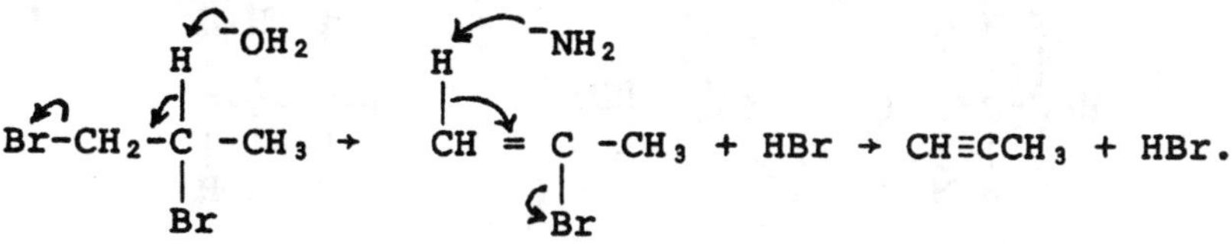

```
         H  -OH2          H    -NH2
         |                |
Br-CH2-C -CH3   →   CH = C -CH3 + HBr → CH≡CCH3 + HBr.
         |                |
         Br               Br
```

(b) propylene: $CH_3CH = CH_2$.

As there is no direct way to change the double bond to a triple bond, it is however possible to produce the 1,2-dibromopropane by brominating in the presence of carbon tetrachloride and then proceeding in the same process outlined in (a).

$$CH_3CH{=}CH_2 + Br_2 \xrightarrow{CCl_4} CH_3CHBrCH_2Br$$

1,2-dibromopropane

(c) isopropyl bromide: $(CH_3)_2CHBr$.

Isopropyl bromide cannot be directly transformed to propyne. Reacting with $NaNH_2$ would not give the desired product:

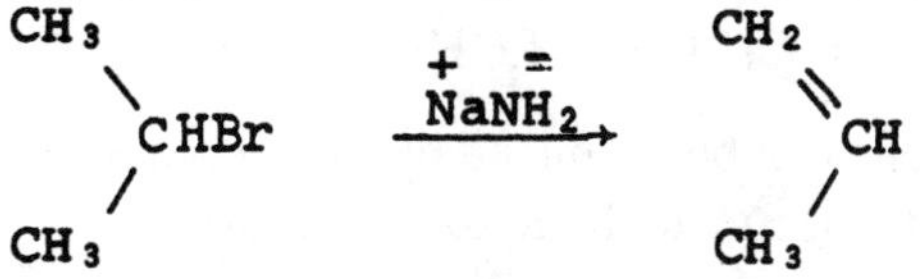

```
CH3                        CH2
   \        +  =              \\
    CHBr   NaNH2  →             CH
   /                           /
CH3                        CH3        Propylene
```

However, propylene can be converted to propyne in the synthesis given in part (b).

$$CH_3CH = CH_2 + Br_2 \rightarrow CH_3CHBrCH_2Br$$

$$CH_3CHBrCH_2Br \xrightarrow{\text{alc KOH}} CH_3-\underset{\underset{Br}{|}}{C}H-CH_2 \xrightarrow[NH_3]{NaNH_2}$$

$$CH_3-C\equiv CH_2$$

(d) propane: C_3H_8.

Since this compound is completely saturated, it can't be halogenated by Br_2 and CCl_4. But by free radical chain reaction, however, we could chlorinate propane.

1-chloropropane

```
  H H H                                 H   H H
  | | |                                 |   | |
H-C-C-C-H   +   Cl2  --hν-->          H-C -C-C-H
  | | |                                 |   | |
  H H H                                 Cl  H H
```

Two isomers are formed:

```
          +        H H   H
                   | |   |
                 H-C-C -C-H
                   | |   |
                   H Cl  H
```

2-chloropropane

If both of these isomers are dehydrohalogenated by potassium hydroxide in alcohol, propylene will result.

$$CH_2Cl-CH_2-CH_3 \xrightarrow[\text{alcohol}]{KOH} CH_2=CH-CH_3$$

Propylene could be converted to propyne by the method outlined in part (b).

(e) n-propyl alcohol: $CH_3CH_2CH_2OH$.

First, propylene can be produced by the dehydration of n-propyl alcohol (to eliminate the hydroxy group).

$$CH_3CH_2CH_2OH \quad + \quad H_2SO_4 \xrightarrow[-\ H_2O]{\text{heat}} CH_3CH=CH_2$$

The hot conc. acid protonates the – OH of the alcohol; subsequent loss of H_2O and H^+ forms the alkene. Propylene can then be converted to propyne as illustrated in part (b).

(f) 1,1–dichloropropane:

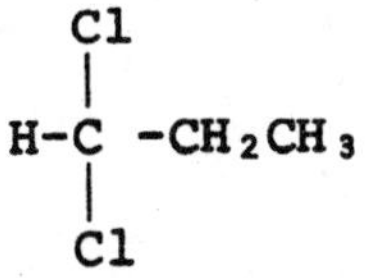

```
      Cl
      |
    H-C -CH2CH3
      |
      Cl
```

By dehydrohalogenation of the dichloropropane we can produce propyne.

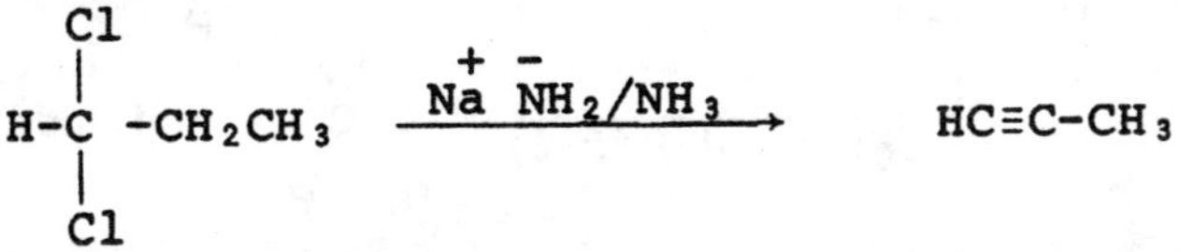

$$H\text{-}C(Cl)_2\text{-}CH_2CH_3 \xrightarrow{\overset{+}{Na}\ \overset{-}{NH_2}/NH_3} HC{\equiv}C\text{-}CH_3$$

(g) Acetylene. Starting from acetylene HC ≡ CH, the carbanion $HC \equiv C^-$ must be obtained. By using $NaNH_2$, which is basic relative to acetylene, the end products are $HC \equiv C^- Na^+ + NH_3$. It is necessary to increase the carbanion by a methyl group to get the desired product, propyne. Since $HC \equiv C^- Na^+$ is fairly nucleophilic, it will participate in an S_N2 displacement reaction with methyl iodide.

$$HC{\equiv}C^{-}\ Na^{+} + CH_3I \rightarrow HC{\equiv}CCH_3 + Na^{+}\ I^{-}$$

(h) 1,1,2,2-tetrabromopropane:

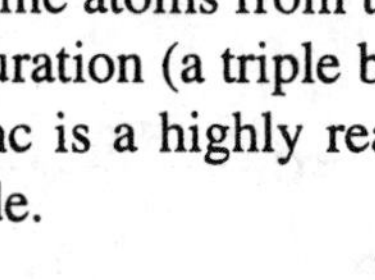

```
      Br Br
      |  |
    H-C -C -CH3
      |  |
      Br Br
```

Instead of dehydrohalogenation, it would be sufficient to remove all the bromine atoms from the molecule and thus form two sites of unsaturation (a triple bond) between the first two carbon atoms. Zinc is a highly reactive element and readily forms zinc bromide.

$$2\,Zn + CHBr_2CBr_2CH_3 \xrightarrow{heat} 2\,ZnBr_2 + HC{\equiv}CCH_3$$

4.4 Reactions of Alkynes

A) Addition of Hydrogen

$$-C \equiv C- \; + 2H_2 \xrightarrow[Pt]{Ni \text{ or}} -\overset{H}{\underset{H}{C}}-\overset{H}{\underset{H}{C}}-$$

$$-C \equiv C- \xrightarrow{Na \text{ or } Li,\, NH_3} \text{(H)C=C(H) trans} \quad \text{trans addition}$$

$$-C \equiv C- \xrightarrow[Pd \text{ or } Ni-B(P-2)]{H_2} \text{(H)C=C(H) cis} \quad \text{cis addition}$$

Example

$$\underset{\text{2-butyne}}{CH_3-C \equiv C-CH_3} + 2H_2 \xrightarrow{Ni} \underset{\text{n-butane}}{CH_3CH_2CH_2CH_3}$$

B) Addition of Halogens

$$-C \equiv C- \xrightarrow{X_2} -\underset{X}{C}=\underset{X}{C}- \xrightarrow{X_2} -\overset{X}{\underset{X}{C}}-\overset{X}{\underset{X}{C}}- \qquad X_2 = Cl_2, Br_2$$

Example

$$CH_3-C \equiv CH \xrightarrow{Br_2} CH_3-\underset{Br}{C}=\underset{Br}{C}H \xrightarrow{Cl_2} CH_3-\overset{Cl}{\underset{Br}{C}}-\overset{Cl}{\underset{Br}{C}}H$$

C) Addition of Hydrogen Halides

$$-C \equiv C- \xrightarrow{HX} -\underset{H}{C}=\underset{X}{C}- \xrightarrow{HX} -\overset{H}{\underset{H}{C}}-\overset{X}{\underset{X}{C}}- \qquad X = Cl, Br, I$$

Example

$$CH_3C \equiv CH \xrightarrow{HCl} CH_3\underset{Cl}{C} = CH_2 \xrightarrow{HI} CH_3-\overset{I}{\underset{Cl}{C}}-CH_3$$

D) Addition of Water (Hydration)

$$-C \equiv C- \quad + \quad H_2O \xrightarrow{H_2SO_4,\, HgSO_4} \underset{\substack{| \quad\ \ | \\ H \quad OH}}{-C = C-} \longrightarrow \underset{\substack{| \quad \| \\ H \quad O}}{-C-C-}$$

more stable form

Example

$$CH_3-C \equiv CH + H_2O \xrightarrow[HgSO_4]{H_2SO_4} H-\underset{\substack{| \\ H}}{\overset{\substack{H \\ |}}{C}}-\underset{\substack{\| \\ O}}{C}-\underset{\substack{| \\ H}}{\overset{\substack{H \\ |}}{C}}-H$$

propyne — 2-propanone (acetone)

E) Nucleophilic Additions

Unlike the simple alkenes, alkynes undergo these additions.

Example

Reaction with alkoxides in alcoholic solution to yield vinyl ethers.

$$CH \equiv CH + RO^- \xrightarrow[\text{High Temp. and Press.}]{ROH} ROCH = CH^- \xrightarrow{ROH} ROCH{=}CH_2 + RO^-$$

Problem Solving Examples:

Give structures and names of the organic products expected from the reaction of 1-butyne with the following:

(a) 1 mole H_2, Ni (b) 2 moles Br_2 (c) 2 moles HCl

A Addition of hydrogen to alkynes leads to the formation of alkenes, if reaction can be limited to the first stage of addition. Two molecules of reagent can be consumed for each triple bond. If the reaction proceeds in two stages, alkanes are formed.

(a) In (a), only one mole of H_2 is added, so that the alkene results.

$$CH_3CH_2C{\equiv}CH \quad + \quad H_2 \xrightarrow{Ni} CH_3CH_2CH{=}CH_2$$

1-butyne — 1-butene

(b) Alkynes react with halogens by addition reactions also. In (b), two moles of the halogen Br_2 react with the triple bond to form a tetrahalide. The reaction scheme appears as:

$$CH_3CH_2C{\equiv}CH + 2Br_2 \rightarrow CH_3CH_2CBr_2{-}CBr_2{-}H$$

1-butyne → 1,1,2,2-tetrabromobutane

(c) Reaction (c) involves the addition of hydrogen halide. The general equation for the reaction is:

$$RC{\equiv}CH \xrightarrow{HX} R{-}C(X){=}CH_2 \xrightarrow{HX} RC(X)_2{-}CH_3$$

Hydrogen chloride, bromide, or iodide adds to alkynes in two steps. The reaction may be stopped after the first step to give the vinyl halide, or a second molecule of HX may be allowed to react to give a gem-dihalide. (The prefix "gem," from the Latin *geminus*, twin, signifies that both halogens are on the same carbon.)

The reaction of 1-butyne with hydrogen chloride follows:

$$CH_3CH_2C{\equiv}CH \xrightarrow{HCl} CH_3CH_2C(Cl){=}CH_2 \xrightarrow{HCl} CH_3CH_2C(Cl)_2{-}CH_3$$

1-butyne → 2-chloro-1-butene → 2,2-dichlorobutane

Describe the reduction of alkynes to obtain an alkene.

A: Reduction of an alkyne to form an alkene is the conversion of the carbon-carbon triple bond of the alkyne to the carbon-carbon double bond of the alkene. Unless the triple bond is at the end of the alkyne, the reaction will yield either a cis-alkene or a trans-alkene. The predominant isomer depends entirely upon the reducing agent used. Examples will make this concept clearer.

$$R-C\equiv C-R \xrightarrow[\text{Pd or Ni-B}]{H_2} \text{cis-}RHC=CHR \quad \text{cis-alkene}$$

$$R-C\equiv C-R \xrightarrow{\text{Na, } NH_3 \text{ (liq.)}} \text{trans-}RHC=CHR \quad \text{trans-alkene}$$

By reduction of alkynes with sodium in liquid ammonia, the trans-alkene will predominate. On the other hand, almost entirely cis-alkene (as high as 98%) is obtained by hydrogenation of alkynes with specially prepared palladium called Lindlar's catalyst or a nickel boride called P-2 catalyst.

Quiz: Alkenes and Alkynes

1. sp bond hybridization for carbon is characteristic in

   (A) alkanes.
   (B) alkenes.
   (C) dienes.
   (D) alkynes.
   (E) aromatics.

2. n-butyl alcohol undergoes dehydration in the presence of 75% H_2SO_4 and 140°F. The chief product of this reaction is

$$CH_3CH_2CH_2CH_2OH \xrightarrow[75\%\ H_2SO_4]{140°F} \text{Product}$$

   (A) $CH_3CH = CHCH_3$
   (B) $CH_2 = CHCH_2CH_3$
   (C) $CH_2OHCH_2CH_2CH_2OH$

(D) $CH_3CHOHCH_2CH_2OH$.

(E) $CH_2 = CHCH = CH_2$.

3. Which of the following is the major product of the reaction below?

Reaction: $C_6H_5CH = CH_2 \xrightarrow[Br_2/CH_3OH]{\text{dilute solution}}$?

(A) $C_6H_5CHBr - CH_2Br$

(B) $C_6H_5\overset{\overset{OCH_3}{|}}{C}HCH_2Br$

(C) $C_6H_5CH_2 - CHBr_2$

(D) $C_6H_5\overset{\overset{OCH_3}{|}}{C}Br - CH_2Br$

(E) None of the above.

4. Which of the following reactions of alkenes is incomplete?

(A) 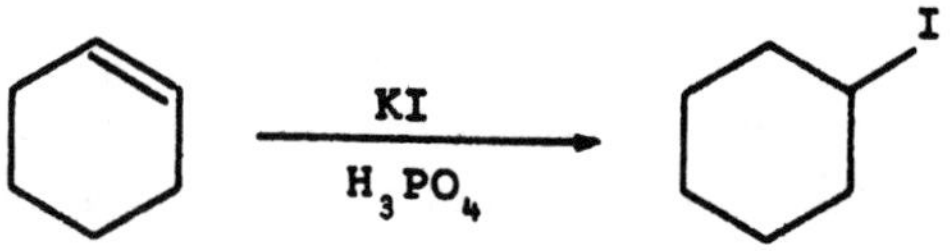 $\xrightarrow[H_3PO_4]{KI}$ I

(B) $(CH_3)_2C = CH_2$ + HCl → $CH_3 - \underset{\underset{Cl}{|}}{\overset{\overset{CH_3}{|}}{C}} - CH_3$

(C) $CH_3CH_2CH = CHCH_3$ + HBr → $CH_3CH_2\overset{\overset{Br}{|}}{C}HCH_2CH_3$

(D) $CH_3\overset{\overset{\displaystyle CH_3}{|}}{C}HCH_2CH = CH_2 + HI \rightarrow CH_3\overset{\overset{\displaystyle CH_3}{|}}{C}HCH_2\overset{\overset{\displaystyle I}{|}}{C}HCH_3$

(E)

$$C_6H_5CH{=}CH_2 \xrightarrow[Br_2/CH_3OH]{\text{dil solution}} C_6H_5CHBr{-}CH_2Br + C_6H_5\overset{\overset{\displaystyle OCH_3}{|}}{C}HCH_2Br$$

5. If butadiene is polymerized by a free radical synthesis, the product contains which repeating units?

(I) $-CH_2(H)C{=}C(H)CH_2-$ (trans: $-CH_2$ and CH_2- on opposite sides; H on each carbon)

(II) $-CH_2(H)C{=}C(H)CH_2-$ ($-CH_2$ upper left, H lower left; H upper right, CH_2- lower right)

(III) $-CH_2-\underset{\underset{\displaystyle CH=CH_2}{|}}{CH}$

(IV) $-CH_2-CH_2-CH_2-CH_2-$

(A) I and II only.

(B) I, II, and III only.

(C) I, II, III, and IV.

(D) III only.

(E) IV only.

6. Alkenes can best be converted to alkynes by way of which intermediate?

(A) ozonide: R–CH ring with O–O and O bridging to C(H)–R

(B) $R-\overset{O}{\overset{\|}{C}}-\overset{O}{\overset{\|}{C}}-R$

(C) $R-\overset{O}{\overset{\|}{C}}-CH_2-R$

(D) $R-CH_2-\overset{Br}{\overset{|}{CH}}-R$

(E) $R-\overset{Br}{\overset{|}{CH}}-\overset{Br}{\overset{|}{CH}}-R$

7. Which of the following reactions will not take place?

(A) $CH_3C\equiv C-CH_3 + H_2 \xrightarrow{Pd}$

(B) $CH_3C\equiv CH + H_2O \xrightarrow[HgSO_4]{H_2SO_4}$

(C) $CH_3-C\equiv C-C_2H_5 + CH_3MgBr \longrightarrow$

(D) $CH_3-C\equiv C-CH_3 + (1)B_2H_6 (2) \xrightarrow[0°C]{CH_3COOH}$

(E) $CH-C\equiv CH + (1)[(CH_3)_2CH\underset{\underset{CH_3}{|}}{CH}]_2BH (2) \xrightarrow[OH^-]{H_2O_2}$

8. Which of the following conditions can produce the reaction below in good yield?

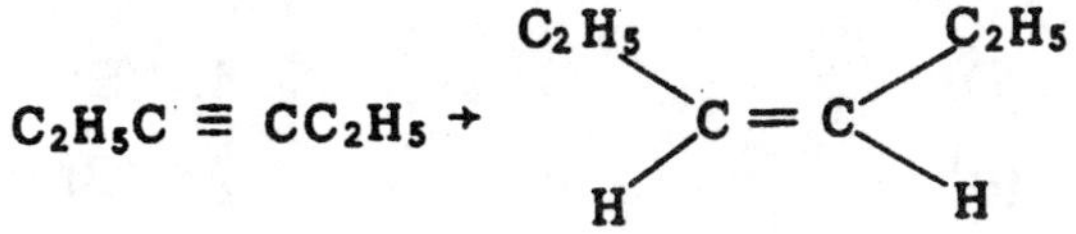

$$C_2H_5C \equiv CC_2H_5 \rightarrow \begin{matrix} C_2H_5 & & C_2H_5 \\ & C=C & \\ H & & H \end{matrix}$$

(A) $\xrightarrow{H_2/Pt}$

(B) $\xrightarrow[\text{liq. } NH_3]{Na} \xrightarrow{NH_4OH}$

(C) $\xrightarrow[\text{quinoline}]{H_2/Pd/BaSO_4}$

(D) $\xrightarrow{KMnO_4}$

(E) $\xrightarrow{HCl}$

9. One would expect the following alkene to be

$$CH_3-CH_2-CH=CH-CH_3$$

(A) a gas.
(B) a liquid.
(C) a solid.
(D) blue in color.
(E) soluble in water.

10. The diene $CH_2=CH-CH=CH_2$ has double bonds that are

(A) conjugate.
(B) sequential.
(C) isolated.
(D) cumulated.
(E) propagated.

ANSWER KEY

1. (D)
2. (A)
3. (B)
4. (C)
5. (B)
6. (E)
7. (C)
8. (C)
9. (B)
10. (A)

CHAPTER 5

Alkyl Halides

Alkyl halides are compounds in which one hydrogen atom is replaced by an atom of the halide family. An important use of alkyl halides is as intermediates in organic synthesis.

Structural formula: $C_nH_{2n+1}X$; X = Cl, Br, I, F.

5.1 Nomenclature (IUPAC System)

Table 5.1

Formula	Name
CH_3Cl	chloromethane
CH_3CH_2Br	bromoethane
$CH_3CH_2CH_2I$	1-iodopropane
CH_3CHICH_3	2-iodopropane
$CH_3CH_2CH_2CH_2Cl$	1-chlorobutane
$CH_3CH_2CHBrCH_3$	2-bromobutane
$(CH_3)_3CI$	2-iodo-2-methylpropane
$CH_3CH_2CH_2CH_2CH_2Cl$	1-chloropentane

5.2 Physical Properties of Alkyl Halides

The specific gravities and boiling points increase with increasing atomic weight of the halogen atom.

Most monohaloalkanes up to C_{18} are liquids at room temperature.

Alkyl halides are usually polar molecules.

Boiling points of alkyl halides are much higher than those of alkanes with the same carbon skeleton.

Alkyl halides are insoluble in water, but are soluble in the typical organic solvents.

They are good solvents for most organic compounds.

They are colorless when pure and have a pleasant odor.

5.3 Preparation of Alkyl Halides

A) From Alcohols

a) Alcohols react with hydrogen chloride or bromide in the presence of sulfuric acid or zinc chloride.

$$R\text{-}OH + HX \xrightarrow{H_2SO_4} R\text{–}X + H_2SO_4 \cdot H_2O$$

R may rearrange

Example

$$CH_3CH_2CH_2OH + HBr \xrightarrow{H_2SO_4} CH_3CH_2CH_2Br + H_2SO_4 \cdot H_2O$$

1-propanol　　　　1-bromopropane

b) Alcohols react with dry hydrogen chloride or dry hydrogen bromide.

$$R - OH + HX, \text{dry} \rightarrow R - X + H_2O$$

Example $CH_3CH_2OH + HBr, \text{dry} \rightarrow CH_3CH_2Br + H_2O$

ethanol　　　　bromoethane

c) Alcohols react with phosphorous halides.

$$R-OH+PX_5 \rightarrow R-X+POX_3+HX$$

and

$$3R-OH+PX_3 \rightarrow 3R-X+H_3PO_3$$

Example $CH_3CH_2OH+PCl_5 \rightarrow CH_3CH_2Cl+POCl_3+HCl$

ethanol chloroethane

d) Alcohols react with thionyl chloride.

$$R-OH+SOCl_2 \rightarrow R-Cl+SO_2+HCl$$

B) Addition of Hydrogen Halides to Unsaturated Hydrocarbons

a) $H_2C=CH_2+HX \rightarrow CH_3-CH_2-X$ primary halide

b) $R-HC=CH_2+HX \rightarrow R-CHX-CH_3$ secondary halide

c) $R_2C=CH_2+HX \rightarrow R_2CX-CH_3$ tertiary halide

C) Halogenation of Alkanes

$$R-H+X_2 \rightarrow R-X+HX$$

Example $CH_4+Cl_2 \rightarrow CH_3Cl+HCl$

chloromethane

D) Halide Exchange

a) $R-X + NaI \xrightarrow{acetone} R-I + NaX \quad X = Cl, Br$

b) $R-X + AgF \longrightarrow R-F + AgX$

c) $R-CCl_3 + SbF_3 \longrightarrow R-C-F_3 + SbCl_3$

Haloform Reaction

Methyl ketones are converted to acids.

$$R-\overset{O}{\overset{\|}{C}}-CH_3 \xrightarrow[\text{or } 3NaOBr \text{ or } 3NaOI]{3NaOCl,\, H_2O} R-\overset{O}{\overset{\|}{C}}-CCl_3 \xrightarrow{NaOH} RCOO^-Na^+ + HCCl_3$$

a tri-chloromethyl ketone; or $HCBr_3$ or HCI_3

Halogenation of Alkenes

Electrophilic addition of X_2 (X = Br,Cl) yields 1,2-dihalides.

$$-\overset{|}{\underset{}{C}}=\overset{|}{\underset{}{C}}- + X_2 \rightarrow -\overset{|}{\underset{\underset{X}{|}}{C}}-\overset{|}{\underset{\underset{X}{|}}{C}}-$$

Hunsdiecker Reaction

Long-chain alkyl bromides are obtained from fatty acids.

$$RCO_2H \rightarrow RBr$$

The resulting alkyl bromide contains one less carbon than the starting compound.

Problem Solving Example:

Describe the Hunsdiecker reaction.

Carboxylic acids undergo several reactions in which the carboxyl group is replaced by halogen.

$$RCO_2H \rightarrow RX$$

Such reactions, in which carbons are lost from a molecule, are called "degradations."

In the Hunsdiecker reaction, the silver salt of a carboxylic acid, prepared by treating the acid with silver oxide, is treated with a halogen. Bromine is the usual reagent, but iodine may also be used. Carbon dioxide is evolved and the corresponding alkyl halide is obtained, usually in fair to good yield.

$$CH_3O\overset{O}{\overset{\|}{C}}CH_2CH_2CH_2CH_2\overset{O}{\overset{\|}{C}}O^-Ag^+ \xrightarrow[CCl_4]{Br_2} CH_3O\overset{O}{\overset{\|}{C}}CH_2CH_2CH_2CH_2Br + AgBr + CO_2$$

The reaction appears to proceed by a free-radical path and may be formulated as follows:

(1) $$R-\overset{O}{\overset{\|}{C}}O^-Ag^+ + Br_2 \longrightarrow R-\overset{O}{\overset{\|}{C}}-OBr + AgBr$$

$$(2)\ R-\overset{O}{\overset{\|}{C}}-OBr \longrightarrow R-\overset{O}{\overset{\|}{C}}-O\cdot + Br\cdot$$

$$(3)\ R-\overset{O}{\overset{\|}{C}}-O\cdot \longrightarrow R\cdot + CO_2$$

$$(4)\ R\cdot + R-\overset{O}{\overset{\|}{C}}-OBr \longrightarrow RBr + R-\overset{O}{\overset{\|}{C}}-O\cdot$$

In a useful modification of the Hunsdiecker reaction, the carboxylic acid is treated with mercuric oxide and bromine.

$$2\ \text{(cyclopropyl)}\begin{smallmatrix}H\\COOH\end{smallmatrix} + HgO + 2Br_2 \rightarrow 2\ \text{(cyclopropyl)}\begin{smallmatrix}H\\Br\end{smallmatrix} + HgBr_2 + 2CO_2 + H_2O$$

(41-46%)

5.4 Reactions of Alkyl Halides

Alkyl halides undergo nucleophilic substitution as shown in Table 5.2.

Reactions of alkyl halides with Grignard reagents produce alkanes.

$$RX + Mg \xrightarrow{\text{dry ether}} RMgX$$

$$R'X + RMgX \longrightarrow R-R' + MgX_2$$

Alkyl halides undergo displacement reactions when treated with aqueous sodium (or potassium) hydroxide, sodium (or potassium) alkoxides, sodium (or potassium) salts of fatty acids, sodium (potassium, silver) cyanide, sodium (potassium, silver) nitrite, and silver oxide.

$$R-X + NaOH,\ aq. \rightarrow R-OH + NaX$$

$$R-X + NaOOCR \rightarrow R-OOCR + NaX$$

$$R-X + NaOR \rightarrow R-OR + NaX$$

$$R-X + NaCN \rightarrow R-CN + NaX$$

$$R-X + AgCN \rightarrow R-CN + AgX$$

$$R-X + NaONO \rightarrow R-ONO + NaX$$

$$R-X + AgO \rightarrow R-OH + AgX$$

Table 5.2 Nucleophilic Substitutions of Alkyl Halides

	Nucleophile		Product	
R−X +	$:\ddot{O}H^-$	Hydroxide	R−OH	Alcohol
	$:\ddot{O}H_2$	Water	R−OH	Alcohol
	$:\ddot{O}R^-$	Alkoxide	R−OR	Ether(Williamson)
	$^-OOC-R'$	Carboxylate	R−OOC−R'	Ester
	$:\ddot{S}H^-$	Hydrosulfide	R−SH	Thiol
	$:\ddot{S}R^-$	Thioalkoxide	R−SR'	Sulfide
	$:\ddot{S}R'_2$	Sulfide	$R-\overset{+}{S}R'_2X^-$	Sulfonium salt
	SCN^-	Thiocyanide	R−SCN	Alkyl thiocyanide
	$:\ddot{I}:^-$	Iodide	R−I	Alkyl iodide
	$:\ddot{N}H_2^-$	Amide	$R-NH_2$	1° Amine
	$:NH_3$	Ammonia	$R-NH_2$	1° Amine
	$:NH_2R'$	1° Amine	R−NHR'	2° Amine
	$:NHR'_2$	2° Amine	$R-NR'_2$	3° Amine
	$:NR_3$	3° Amine	$R-\overset{+}{N}R'_3X^-$	Quaternary ammonium salt
	N_3^-	Azide	$R-N_3$	Alkyl azide
	NO_2^-	Nitrite	$R-NO_2$	Nitroalkane
	$:P(C_6H_5)_3$	Phosphine	$R-\overset{+}{P}(C_6H_5)_3X^-$	Phosphonium salt
	$^-:C\equiv N:$	Cyanide	R−CN	Nitrile
	$^-:C\equiv C-R'$	Alkynyl anion	$R-C\equiv C-R'$	Alkyne
	$^-:R'$	Carbanion	R−R'	Alkane
	$^-:CH(COOR')_2$		$R-CH(COOR')_2$	Malonic ester synthesis
	$^-:\ddot{C}H(COCH_2)(COOR)$		$R-CH(COCH_2)(COOR)$	Acetoacetic ester synthesis
	$Ar-H, AlCl_3$		R−Ar	Alkyl benzene (Friedel-Crafts)

Problem Solving Example:

Propose a synthesis for n-butane from n-butylbromide using the Grignard reaction.

The Grignard reaction is a common method for synthesizing alkanes from alkyl halides. The general equations for this reaction are as follows:

(i) $R - X \xrightarrow{Mg} RMgX$

(ii) $RMgX \xrightarrow{H_2O} R - H + MgX_2$

The Grignard reagent (RMgX) is formed by treating n-butylbromide with clean magnesium in dry ether.

(iii) $CH_3(CH_2)_2CH_2\text{-}Br + Mg \xrightarrow[\text{ether}]{\text{dry}} CH_3(CH_2)_2CH_2MgBr$

The desired product, n-butane, can then be synthesized by adding H_2O to the Grignard reagent.

(iv) $CH_3(CH_2)_2CH_2MgBr \xrightarrow[H_2O]{} \underset{\text{n-butane}}{CH_3(CH_2)_2CH_3} + MgBr_2$

The intermediate reaction of reaction (iv) is

(v) $CH_3(CH_2)_2CH_2MgBr + HOH \rightarrow CH_3(CH_2)_2CH_3 + MgBrOH$

5.5 Nucleophilic Displacement Reactions

Alkyl halides undergo nucleophilic substitution (S_N) and elimination (E) in the presence of basic reagents. The two reactions are always in competition.

```
                    H
                    |
                  -C-C-
                    | |
Substitution (S_N)    X    Elimination (E)
+ :B⁻                      + :B⁻             :B⁻ = nucleophile
                                                   (basic re-
  H                                                 actant
  | |
-C-C-   + :X⁻             >C=C<  + H:B + :X⁻   :X⁻ = dis-
  | |                                              placed halide
    B                                              (weaker base)
```

In the S_N reaction the base, : B^-, replaces the weaker base, : X^-. The *E* reaction is the reverse of an addition reaction, and a hydrogen and a halogen on adjacent carbons are eliminated. Typical bases were shown in Table 5.2.

Problem Solving Examples:

Q When isopropyl bromide is treated with sodium ethoxide in ethanol, propylene and ethyls isopropyl ether are formed in a 3:1 ratio. If the hexadeuteroisopropyl bromide, $CD_3CHBrCD_3$, is used, $CD_3CH = CD_2$ and $(CD_3)_2CHOC_2H_5$ are formed in a ratio of 1:2. Explain.

A Isopropyl bromide is an alkyl halide. To solve this problem, it will be necessary to consider the typical reactions of alkyl halides.

The halide ion may be characterized as a weak base. It can readily be displaced by stronger bases that possess an unshared pair of electrons and seek a relatively positive site. Such basic, electron-rich reagents are nucleophilic reagents. Consequently, one typical reaction of alkyl halides is nucleophilic substitution:

```
    R : X          +      : Z                  R : Z + : X⁻
(Alkyl halide) (Nucleophilic Reagent)  ——→   (leaving group)
```

One type of nucleophilic substitution is called S_N2, which stands for bimolecular nucleophilic substitution. Bimolecular deals with the kinetics of the reaction. It indicates that the rate determining step involves collision of two particles. For example:

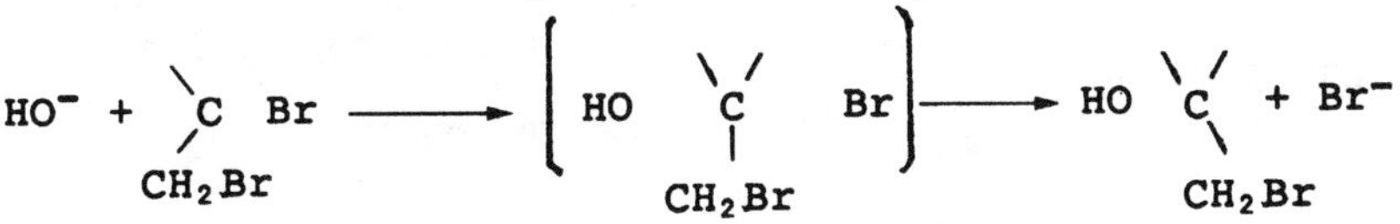

```
     \                      [        \ /          ]             \ /
HO⁻ + C  Br  ———————→       [ HO      C        Br ]———→ HO       C   + Br⁻
     /                      [         |           ]               \
   CH₂Br                            CH₂Br                        CH₂Br
```

In S_N2 reactions, the order of reactivity of RX is $CH_3X > 1° > 2° > 3°$. As the number and size of substituents attached to the carbon bearing the halogen is increased, the reactivity toward S_N2 decreases.

Another typical reaction of alkyl halides is bimolecular elimination (called E_2) to produce alkenes. A base abstracts a hydrogen ion away from carbon, and simultaneously a halide ion separates. It is depicted below:

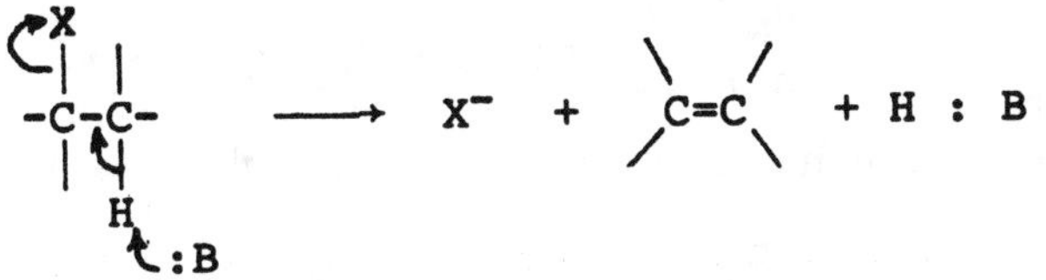

```
   X
   |  |                            \     /
  -C--C-     ———→   X⁻  +          C=C        + H : B
   |  |                            /     \
      H
       :B
```

This reaction has an isotope effect in that deuterium-carbon bonds are broken more slowly than C – H bonds. The order of reactivity of alkyl halides toward E_2 elimination is $3° > 2° > 1°$. It reflects the relative stabilities of the alkenes being formed.

The structure of isopropyl bromide may be written as

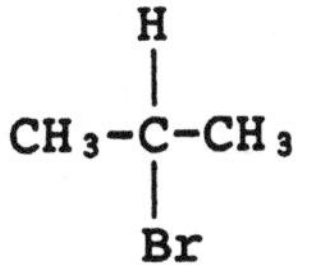

```
       H
       |
  CH₃-C-CH₃
       |
       Br
```

Hence, this alkyl halide is secondary (2°). This means that in the presence of the strong base sodium ethoxide in ethanol, both the S_N2 and E_2 reactions occur; that is, they compete with each other. This explains why two products, propylene and ethyl isopropyl ether, are obtained.

$$CH_3-\underset{Br}{\overset{H}{C}}-CH_3 \xrightarrow{S_N2} CH_3\overset{OC_2H_5}{CH}CH_3 \text{ (ethyl isopropyl ether)}$$

$$CH_3-\underset{Br}{\overset{H}{C}}-CH_3 \xrightarrow{E_2} CH_3-CH=CH_2 \text{ (propylene)}$$

The ratio of the alkene to the ether is 3:1.

With hexadeuteroisopropyl bromide, $CD_3CHBrCD_3$, the ratio changes to 1:2. This is to be expected for the loss of deuterium in E_2 is slower than hydrogen (the isotope effect). The S_N2 reaction that produces the ether begins to predominate. It may be pictured in the following manner:

$$CD_3\overset{OC_2H_5}{CH}CD_3 \xleftarrow[\text{same rate}]{S_N2} CD_3CHBrCD_3 \xrightarrow[\text{slower}]{E_2} CD_3CH=CD_2$$

The deuterium isotope effect may be calculated as shown:

$$\frac{K_D}{K_H}(E_2) = \frac{1/2}{3/1} = \frac{1}{6}$$

Q Predict the order of reactivity of the following halides with (a) sodium iodide in acetone and (b) aqueous alcoholic silver nitrate.

[cis-1-chloro-2-methylcyclopentane: ring with wedged CH_3 and hashed Cl] [(chloromethyl)cyclopentane: ring–CH_2Cl] [1-chloro-1-methylcyclopentane: ring bearing CH_3 and Cl on the same carbon]

$CH_3CH_2CH_2CH=CHCH_2Cl$

A To solve this problem, we consider various techniques used in figuring the relative reactivities of several compounds in S_N reactions.

The first step in solving this problem systematically is to consider the reaction conditions. In part (a) the reaction conditions are sodium iodide in acetone. This media favors S_N2 over S_N1 reactions. In part (b) the reaction conditions are aqueous alcoholic silver nitrate. This indicates an S_N1 type media.

We now consider what factors affect reactivity in S_N reactions. The most important factor is the degree of alkylation of the carbon α to the halogen (the carbon bonded to the halide), that is, whether this alkyl halide is 1°, 2°, or 3°. For review a 1° (primary) alkyl halide follows the form

```
  R
  |
H-C-H,
  |
  X
```

a 2° (secondary) alkyl halide follows the form

```
  H
  |
R-C-R
  |
  X
```

and a 3° (tertiary) alkyl halide follows the form

```
  R
  |
R-C-R
  |
  X
```

where X represents a halogen and R represents an alkyl or aromatic group.

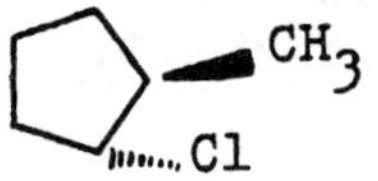

CH_3
Cl

is thus a 2° alkyl halide,

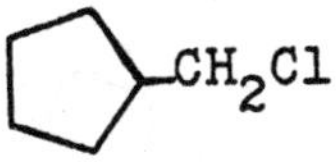

CH_2Cl

is a 1° alkyl halide, and

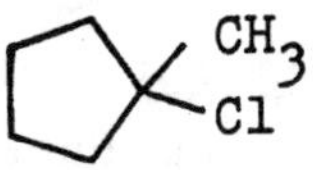

CH_3
Cl

is a 3° alkyl halide. $CH_3CH_2CH_2CH = CHCH_2Cl$ is a 1° alkyl halide but it is a special case in both S_N1 and S_N2 reactions. In part (a) $CH_3CH_2CH_2CH = CHCH_2Cl$ undergoes an S_N2 reaction as follows: $CH_3CH_2CH_2CH = CHCH_2Cl + I^- \xrightarrow{\text{acetone}} CH_3CH_2CH_2CH = CHCH_2I + Cl^-$. The following activated complex is formed in this S_N2 reaction.

$$CH_3CH_2CH_2CH=CHCH_2Cl + I^- \xrightarrow{\text{acetone}}$$

π overlap stabilizing the S_N2 activated complex:

$$\longrightarrow C_3H_8CH=CHCH_2I + Cl^-$$

This activated complex has π bonding overlap, which tends to stabilize and thus decrease the energy content of the activated complex. The net result of all this is that $C_3H_8CH = CHCH_2Cl$ (1°) will have a greater reactivity than

because of this effect.

Getting back to the degree of the alkyl halide, in an S_N2 reaction the order of reactivity is $CH_3X > 1° > 2° > 3°$, while in a S_N1 reaction the order of reactivity is $3° > 2° > 1° > CH_3X$. Thus, in part (a) (S_N2 media) the order of reactivity is

$C_3H_8CH = CHCH_2Cl >$

In part (b) (S_N1 media) $C_3H_8CH = CHCH_2Cl$ exhibits another very interesting effect. You will note that this compound has an allylic type group:

The compound reacts via an S_N1 path in the following way:

$$C_3H_8CH=CHCH_2Cl \xrightarrow{-Cl^-} C_3H_8CH=CH\overset{+}{C}H_2 \xrightarrow{I^-} C_3H_8CH=CHCH_2I$$

allylic carbocation intermediate

The allylic carbocation intermediate displays what is called a resonance effect. That is, the allylic carbocation can be drawn as two identical arrangements of atoms, differing only in their electronic distribution. Each arrangement is called a resonance or contributing structure. These structures have no physical existence and they are not in equilibrium; rather the real allyl cation is a hybrid of the two main contributing structures (hybrid also in electronic distribution):

$$C_3H_8(H)C{=}CH{-}C^{+}H_2 \longleftrightarrow C_3H_8(H)C^{+}{-}CH{=}CH_2$$

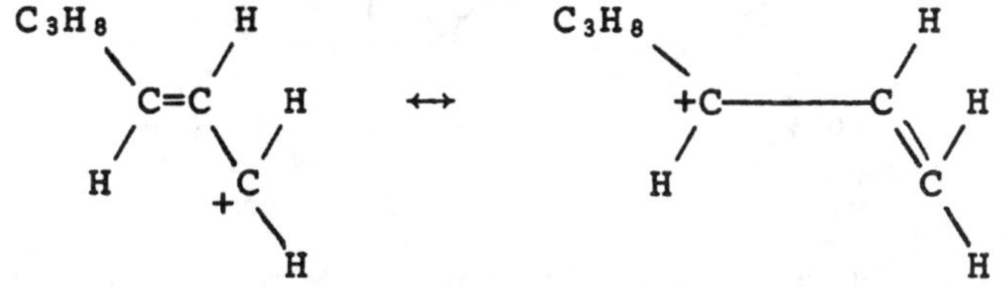

Figure 5.1

Notice from Figure 5.1 that the positive charge is delocalized among the contributing structures. This delocalization of charge greatly stabilizes the carbocation, thus creating an effect whereby $C_3H_8CH = CHCH_2Cl$ is more reactive via S_N1 than even t-butyl bromide, a tertiary compound.

Taking this resonance effect into account, the order of reactivity in part (b) is:

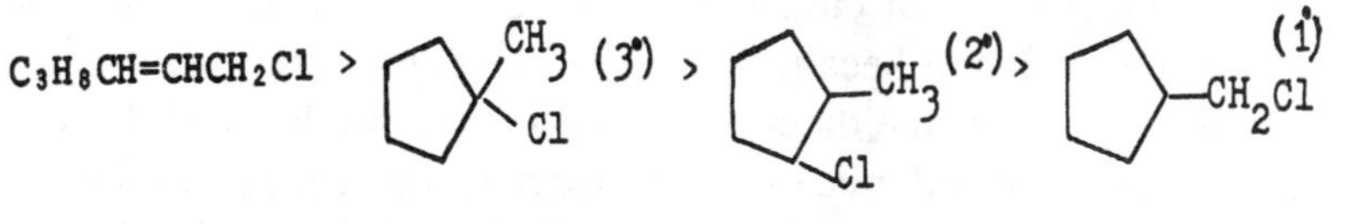

$C_3H_8CH{=}CHCH_2Cl$ > 1-chloro-1-methylcyclopentane (CH_3, Cl) (3°) > 1-chloro-2-methylcyclopentane (CH_3, Cl) (2°) > chloromethylcyclopentane ($-CH_2Cl$) (1°)

5.6 S_N1 and S_N2 Substitution Reactions

A nucleophilic substitution reaction (S_N reaction) is the typical reaction encountered by an alkyl halide in the presence of a basic electron-rich reagent (nucleophile).

$$R{:}X + {:}B^- \rightarrow R{:}B + {:}X^-$$

Example $\quad CH_3Br + {:}OH^- \rightarrow CH_3OH + {:}Br^-$

The mechanism of the above bimolecular S_N reaction (S_N2) involves a direct collision between the nucleophile ($:OH^-$) and the carbon bearing the halide. The S_N2 reaction follows second order kinetics because its rate depends upon the concentrations of the two reacting substances.

Inversion of configuration has taken place when a reaction yields a product whose configuration is opposite to that of the reactant. The S_N2 reaction causes complete stereochemical inversion due to backside attack.

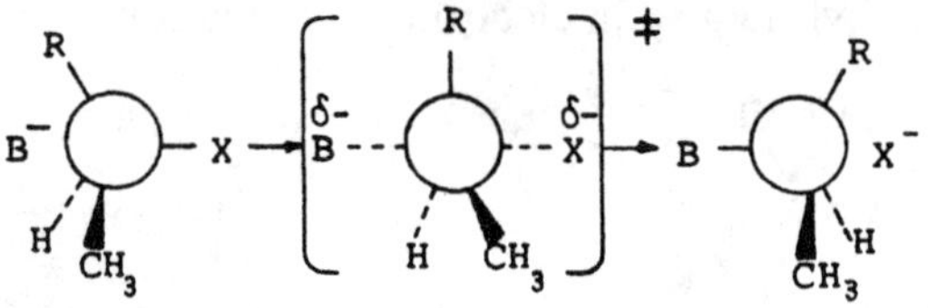

S_N2: complete inversion

The reactivity toward S_N2 substitution decreases as the number of substituents attached to the carbon carrying the halogen increases. The order of reactivity is

$$CH_3X > 1° > 2° > 3°$$

Another example of a nucleophilic substitution reaction is:

$$(CH_3)_3C{-}Br + :OH^- \rightarrow (CH_3)_3C{-}OH + :Br^-$$

The mechanism of the above unimolecular S_N reaction (S_N1) does not involve collision because the rate determining step involves only one molecule. The rate determining step is the single step whose rate determines the overall rate of a stepwise reaction. The rate of the entire reaction is determined by how fast the alkyl halide ionizes, and it therefore depends upon only the concentration of the alkyl halide. As a result the S_N1 reaction follows first-order kinetics.

The S_N1 reaction is characterized by the formation of a carbonium ion. The carbonium ion may be attacked by the hydroxide ion from either side of its plane so as to cause either an inversion or retention of configuration.

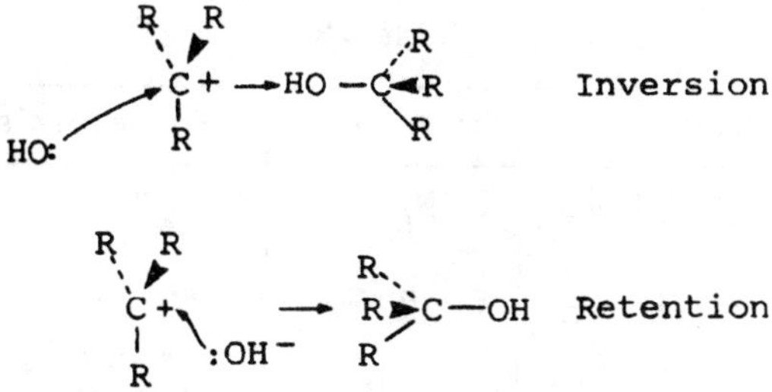

HO: → R_3C^+ → $HO-CR_3$ Inversion

R_3C^+ + :OH⁻ → R_3C-OH Retention

S_N1: Inversion and Retention

In contrast to an S_N2 reaction, the S_N1 reaction proceeds with racemization (both inversion and retention of configuration).

In S_N1 reactions the order of reactivity is allyl, benzyl > 3° > 2° > 1° > CH_3X.

The rate of S_N2 reactions is affected by steric factors (the bulk of the substituents). The rate of S_N1 reactions is affected by electronic factors (the tendency of substituents to release or withdraw electrons). Electron-donating groups favor the S_N1 mechanism, in which a positive charge is generated. Electron-withdrawing groups favor the S_N2 mechanism, in which the transition state is more negative than the starting material.

High concentrations of the nucleophilic reagent favor S_N2 reaction; low concentrations favor S_N1 reaction.

The polarity of the solvent also determines the mechanism by which the reaction occurs. Increasing solvent polarity favors the S_N1 reaction (slows down the S_N2).

Table 5.3

Compound	Structure	S_N Mechanism
Primary Alkyl Halides	RCH_2-X	S_N2
Secondary Alkyl Halides	R_2CH-X	Either S_N1 or S_N2, depending upon the solvent present.
Tertiary Alkyl Halides	R_3C-X	S_N1 occurs when an ionizing solvent is present. A very slow S_N2 occurs when a non-ionizing solvent is used.

The effects of the nucleophile on reaction rate are as follows:

A) The nucleophilicity for a given atom increases with an increase in the negative charges: $:HO^-$ is more nucleophilic than H_2O.

B) An increase in the atomic number within a row of the periodic table decreases nucleophilicity: $C^- > N^- > O^- > F^-$, or $NH_3 > H_2O > HF$.

C) An increase in the atomic number within a column of the periodic table increases nucleophilicity: $I^- > Br^- > Cl^- > F^-$.

Problem Solving Examples:

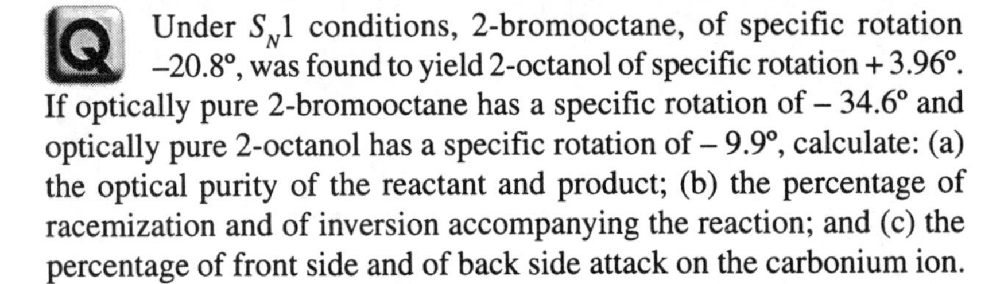

Q Under S_N1 conditions, 2-bromooctane, of specific rotation –20.8°, was found to yield 2-octanol of specific rotation + 3.96°. If optically pure 2-bromooctane has a specific rotation of – 34.6° and optically pure 2-octanol has a specific rotation of – 9.9°, calculate: (a) the optical purity of the reactant and product; (b) the percentage of racemization and of inversion accompanying the reaction; and (c) the percentage of front side and of back side attack on the carbonium ion.

(a) The optical purity of a compound can be expressed as the fraction

$$\frac{r_i}{r_p},$$

where r_i is the rotation of the impure compound and r_p is the rotation of the pure compound. In the case of 2-bromooctane, the optical purity is

$$\frac{20.8}{34.6} \times 100 = 60\%$$

In the case of 2-octanol,

$$\frac{3.96}{9.90} \times 100 = 40\%$$

(b) The percentage of racemization and inversion can be obtained by first determining the ratio of the optical purity of the products to that of the reactants. In the case of 2-bromooctane and 2-octanol, this ratio is

$$\frac{40}{60} \text{ or } \frac{2}{3}.$$

If the reaction had proceeded with complete inversion, an optical purity ratio of 1 would have been expected (for every molecule of 2-bromooctane, one molecule of 2-octanol would have formed).

Therefore, if the optical purity ratio is

$$\frac{2}{3},$$

the reaction proceeds with

$$\frac{2}{3}$$

inversion and

$$\frac{1}{3}$$

racemization.

(c) Every molecule that undergoes front-side attack (retention) cancels the optical activity of a molecule that is undergoing back-side attack (inversion). In this reaction,

$$\frac{1}{6}$$

front-side attack occurs, cancelling

$$\frac{1}{6}$$

of the backside attack, giving

$$\frac{1}{6}+\frac{1}{6}$$

or

$$\frac{1}{3}$$

racemization. Therefore,

$$\frac{1}{6}+\frac{2}{3}$$

(inversion) or

$$\frac{5}{6}$$

of the molecules under backside attack.

Q (a) What product would be formed if the reaction of cis-4-bromocyclohexanol with OH^- proceeded with inversion? (b) Without inversion? (c) Is it always necessary to use optically active compounds to study the stereochemistry of substitution reactions?

(a) If the S_N2 reaction of cis-4-bromocyclohexanol proceeded with inversion, trans (1,4) cyclohexadiol would be produced.

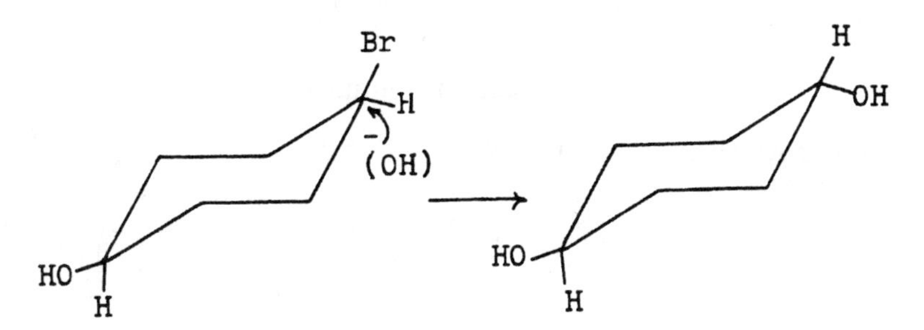

(b) Without inversion (retention):

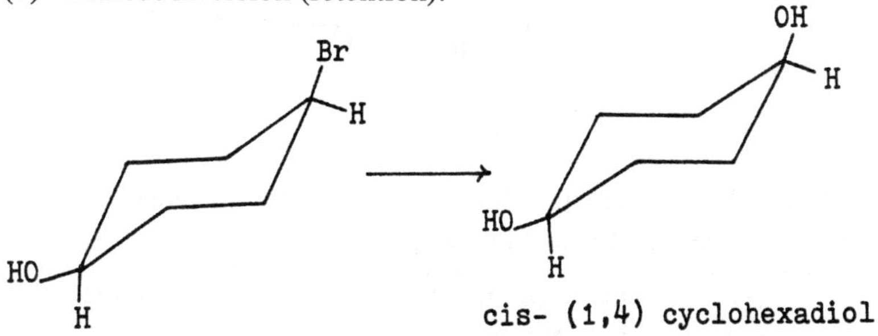

(c) It is not necessary to use optically active reactants to study the stereochemistry of substitution reactions. We may use starting materials such as those in this problem, which have two chiral centers. Using the proper nucleophile, we can induce the reactants to form diastereomeric products which can be separated by differences in physical properties.

5.7 Elimination Reaction: *E*2 and *E*1

*E*2

Bimolecular Elimination

X

–C–C– ⟶ $:X^-$ + >C=C< + H:B

H

$:B^-$

*E*1

Unimolecular Elimination

Rate determining step

(1) $-\overset{X}{\underset{|}{C}}-\underset{H}{C}- \longrightarrow :X^- + -\overset{+}{C}-\underset{H}{C}-$ slow

(2) $-\overset{+}{C}-\underset{H}{C}- \longrightarrow \;\rangle C=C\langle\; + H:B$ fast

:B⁻

The *E*2 mechanism (elimination, bimolecular) involves two molecules in the rate-determining step.

The *E*l mechanism (elimination, unimolecular) involves one molecule in the rate-determining step.

Reactivity toward 3° > 2° > 1°

*E*2 or *E*1 Elimination

The *E*1 elimination mechanism follows first-order kinetics, demonstrates the identical effect of the structure reactivity, and can be accompanied by rearrangement of configuration due to the intermediate carleonium ion formed.

The *E*2 elimination mechanism follows second-order kinetics, is not accompanied by rearrangements. It demonstrates a large deuterium isotope effect, does not undergo hydrogen-deuterium exchange, and demonstrates large element effects.

There is a variable *E*2 elimination mechanism involving the formation of a carbanion.

(1) $-\overset{X}{\underset{|}{\overset{|}{C}}}-\overset{|}{\underset{H}{C}}- \quad :B^- \quad \longrightarrow \quad -\overset{X}{\underset{|}{\overset{|}{C}}}-\overset{|}{\underset{-}{\ddot{C}}}- + H:B$

carbanion

$-\overset{X}{\underset{|}{\overset{|}{C}}}-\overset{|}{\underset{-}{\ddot{C}}}- \longrightarrow :X^- + \; \rangle C{=}C \langle$

Variable *E*2 Mechanism

The mechanism that will be operative in a given situation depends primarily on the structure of the substrate, the nature of the solvent, the strength of the base, and the nature of the leaving group.

The ionization of a carbon-hydrogen bond occurs more readily if the carbon atom is highly branched. Ionization proceeds only if an ionizing solvent is present.

The order of decreasing reactivity in ionization is

tertiary > secondary > primary > methyl, vinyl

Problem Solving Examples:

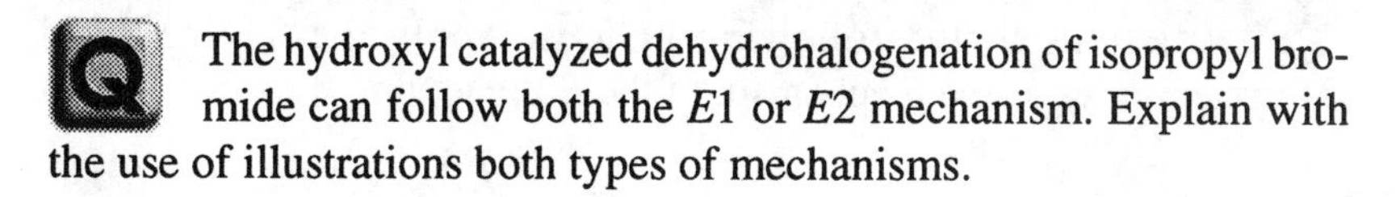

Q The hydroxyl catalyzed dehydrohalogenation of isopropyl bromide can follow both the *E*1 or *E*2 mechanism. Explain with the use of illustrations both types of mechanisms.

A Elimination reactions may be classified as eliminations (eliminations of two groups from the same atom) or eliminations (eliminations of two groups from adjacent atoms yielding unsaturated bonds). Both of these eliminations are catalyzed by bases.

Bimolecular elimination, known as *E*2, involves the abstraction of a β proton by base attack with the simultaneous departing of the group X. It can be illustrated as follows:

$B:^- \quad -\overset{|}{\underset{H}{\overset{\beta}{C}}}-\overset{X}{\underset{|}{\overset{\alpha}{C}}}- \rightarrow B{:}H + \rangle C{=}C \langle + :X^-$

Base

The reaction follows second-order kinetics; the rate of alkene formation depends on the concentrations of both reactants. No rearrangements of the alkyl group are involved in the *E*2 mechanism.

Unimolecular elimination, known as *E*1, differs from *E*2 in that the group X leaves the molecule before the attack of base. As a result, one has an intermediate formation of a carbonium ion. The two-step reaction of *E*1 can be written as:

Step 1: Carbonium ion formation

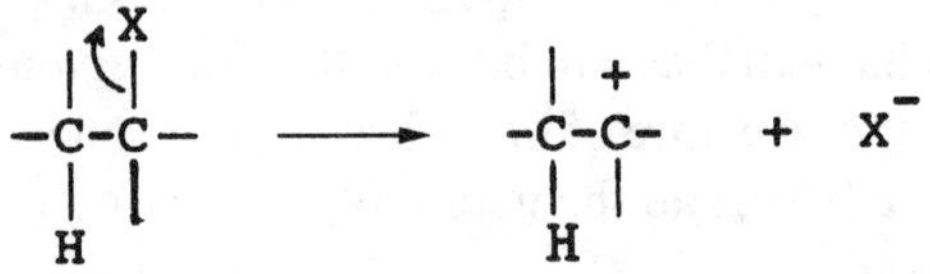

$$\text{H}-\overset{|}{\underset{|}{\text{C}}}-\overset{|}{\underset{|}{\text{C}}}-\text{X} \longrightarrow \text{H}-\overset{|}{\underset{|}{\text{C}}}-\overset{+}{\underset{|}{\text{C}}}- \; + \; X^-$$

Step 2: Abstraction of proton by base attack

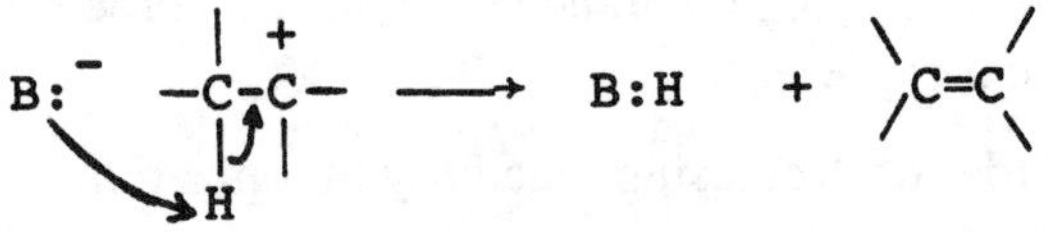

$$B:^- \; + \; \text{H}-\overset{|}{\underset{|}{\text{C}}}-\overset{+}{\underset{|}{\text{C}}}- \longrightarrow B:H \; + \; >C{=}C<$$

Step 1 is the slower and, hence, the rate-controlling step, dependent only on the concentration of the alkyl halide.

Q Predict the structure of the major alkene formed in the *E*2 reactions of the following halides. Apply the Saytzeff and Hofmann rules to the mechanism of the *E*2 reaction.

(a) $CH_3CH_2C(CH_3)(Br)CH_3$ (b) 1-chloro-1-methylcyclohexane (cyclohexane ring bearing CH_3 and Cl on the same carbon)

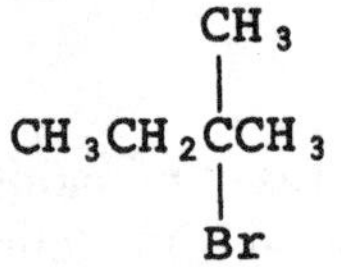

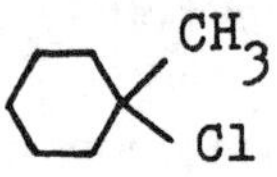

A This problem represents an exercise in applying the Saytzeff and Hofmann rules to the mechanism of an *E*2 reaction. To understand these rules, we will go through the *E*2 mechanism.

The Saytzeff rule predicts that in *E*2 elimination, the product with the most isomer will predominate. In the Hofmann orientation, the less substituted alkene will predominate as product.

An *E*2 will be favored over an S_N2 reaction in conditions such as low polar media (an alcoholic solvent is generally used), high temperature, and when a strong base is present. Mechanism (general case): Note: → represents flow of electrons.

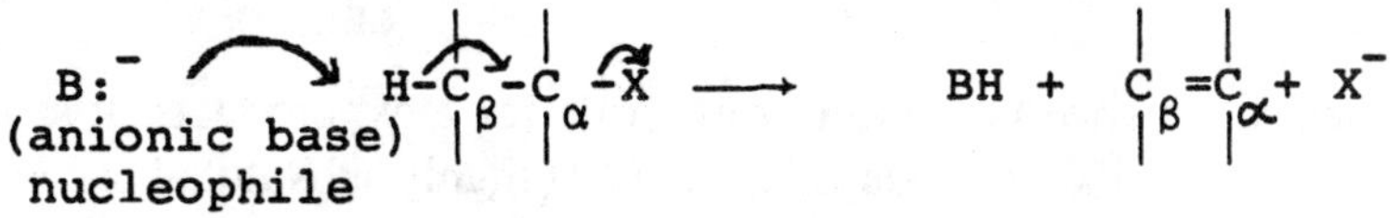

```
B:⁻  ⌒↘  H-Cβ-Cα-X  ⟶   BH + Cβ=Cα + X⁻
(anionic base)
 nucleophile
```

This reaction, which like an S_N2 is concerted (occurs in one step), has the same activated complex as an S_N2.

```
activated         δ-                          δ-
complex in E2     B ----H----C----C----X
```

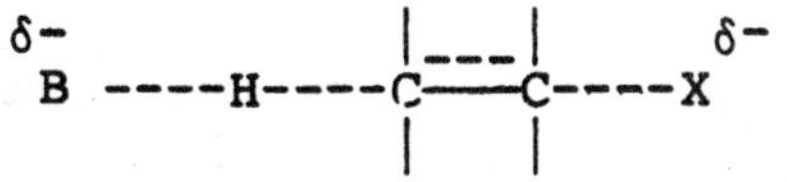

The activated complex shows the concerted process of bond making and bond breaking. In the *E*2 reaction, a nucleophile (B:–) attacks a hydrogen β to the leaving group (–X:). While this occurs, the electron pair of the H-Cβ bond "swings around" and attacks the carbon (Cα) adjacent to it. Cα, unable to hold 10 electrons will simultaneously begin to eject the leaving group along with its two electrons of the Cα –X bond. As the leaving group is ejected, a π bond forms between Cα and Cβ, creating a double bond. Note that this process occurs in one continuous step (concerted). A problem arises when there is more than one type of β carbon that bears a hydrogen. For example, in part (a) we are given the compound

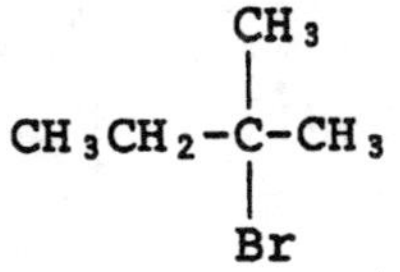

```
          CH3
          |
  CH3CH2-C-CH3
          |
          Br
```

and we are asked to predict the major structure of the alkene formed by an *E*2 reaction.

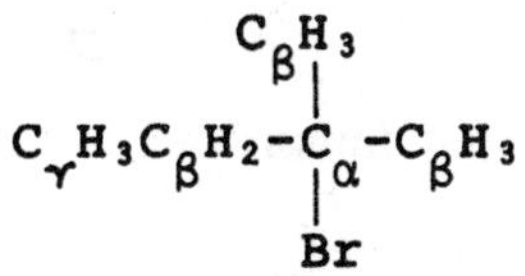

```
            CβH3
            |
  CγH3CβH2-Cα-CβH3
            |
            Br
```

One can readily see that the two β-methyl groups have no difference stereochemically; therefore, the question is which alkene will form:

```
         CH3                      CH2
        /                        //
CH3CH=C          or  CH3CH2-C
        \                        \
         CH3                      CH3
```

To answer this we follow Saytzeff's rule for most products. It states that the major alkene produced is the most highly substituted one (i.e., the one with the largest number of *alkyl* groups bonded to the sp^2- hybridized carbons). In part (a) we readily see that

```
             CH3
            /(2)
   CH3CH =C
(1)         \
             CH3
           (3)
```

has three substituents bonded to the Csp^2 atoms, while with

```
                CH2
               //
(2) CH3CH2-C
               \
                CH3
              (1)
```

we observe only two substituents. According to Saytzeff,

```
         CH3
        /
CH3CH=C
        \
         CH3
```

is the major alkene compound formed, and it is thus known as the Saytzeff product while

```
           CH2
          //
CH3CH2-C
          \
           CH3
```

is the less substituted and minor alkene formed and is called the Hofmann product.

Hofmann's rule states that the major alkene produced is the least highly substituted one (i.e., the one with the smallest number of alkyl groups bonded to the sp^2-hybridized carbons). Hofmann's rule will usually apply when the leaving group is a positively charged species (e.g., $-\overset{+}{N}(CH_3)_3$.

In part (b) we consider which alkene will be produced upon an *E*2 reaction involving

CH_3
Cl

The major compound will either be

(1) CH_2 (2) or (1) (2) CH_3 (3)

Here, Saytzeff's rule applies since the leaving group is Cl^-, a negatively charged species. Thus, the major alkene formed is

CH_3

5.8 Stereochemistry of Elimination

The *E*2 elimination yields two pairs of enantiomers, cis and trans. This elimination mechanism is stereospecific.

Example

C_6H_5 / CH_3—H / Br—H / C_6H_5 I or C_6H_5 / H—CH_3 / H—Br / C_6H_5 II → CH_3 C_6H_5 C=C H C_6H_5

1-bromo-1,2-diphenylpropane cis-1,2-diphenyl1-propene

C_6H_5 / CH_3—H / H—Br / C_6H_5 III or C_6H_5 / H—CH_3 / Br—H / C_6H_5 IV → CH_3 C_6H_5 C=C C_6H_5 H

1-bromo-1,2-diphenylpropane trans-1,2-diphenyl-1-propene

Anti-elimination is involved in the bimolecular reaction of alkyl halides. The hydrogen and the leaving group, in its transition state, are far apart from each other.

The anti-relationship is shown as follows:

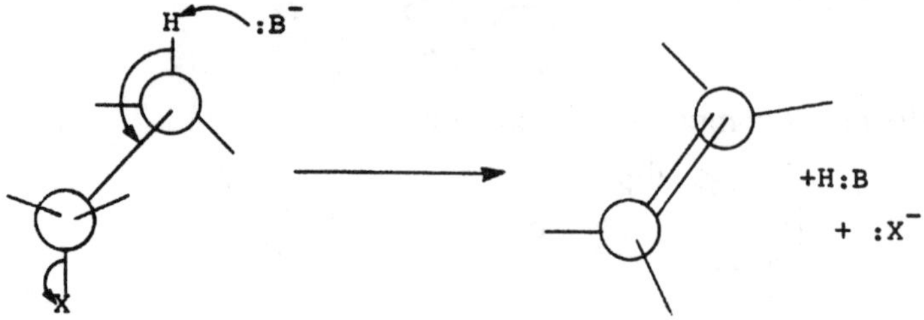

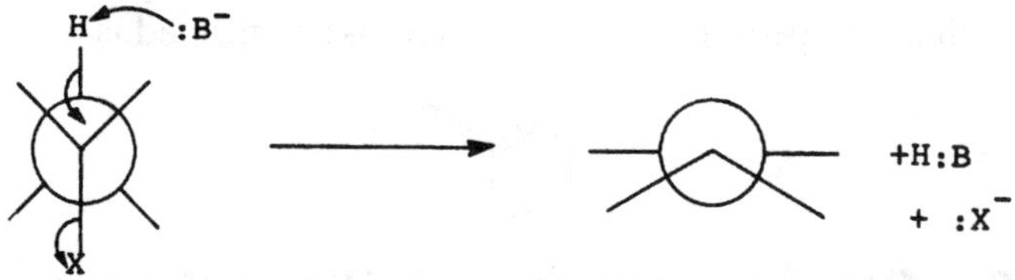

Syn-elimination is also involved in the *E*2 reactions. The hydrogen and the leaving group, in the transition state, are in the eclipsed (or gauche) relationship.

The anti-elimination mechanism is the more stable mechanism, because the anti-conformation is far more stable than the eclipsed conformation.

Problem Solving Examples:

Write all the possible staggered conformations for each of the isomers of 2,3-dibromobutane shown in (1) and (2).

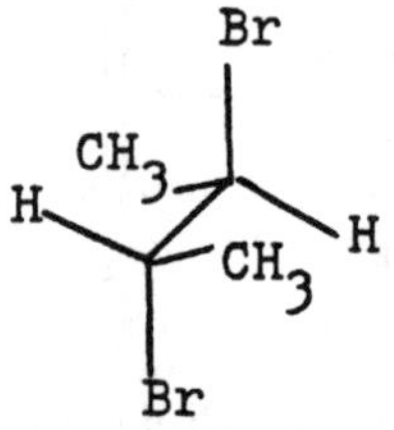

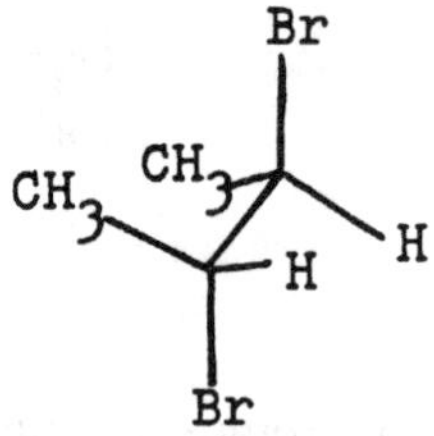

Show the structures of the alkenes that could be formed from each by a trans *E*2 elimination of one mole of hydrogen bromide with hydroxide ion. Which alkene should more readily eliminate further to form dimethylacetylene? Explain.

A A staggered conformation is defined as a conformation with a torsional angle of 60° as seen in a Newman projection. Below are the Newman projections for (1) and (2).

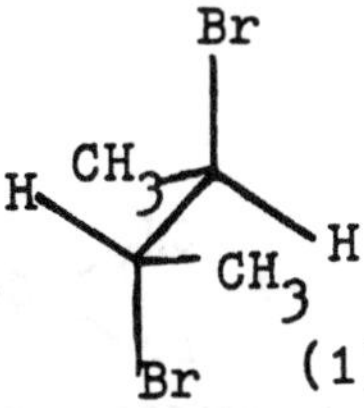

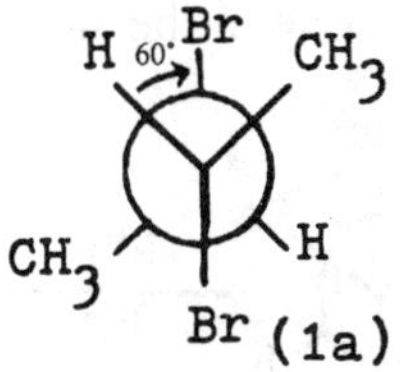

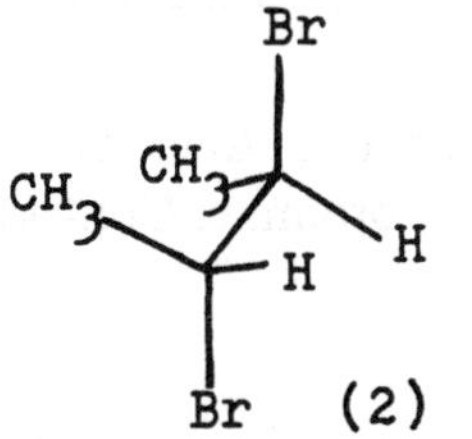

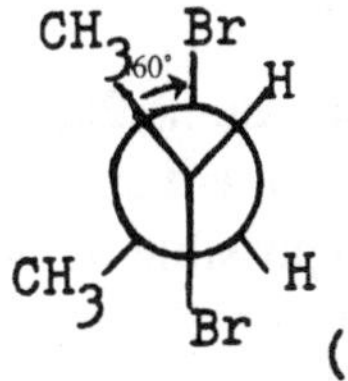

For (1) and (2) the only other Newman projections with torsional angles of 60° and thus the only other staggered conformations are shown below:

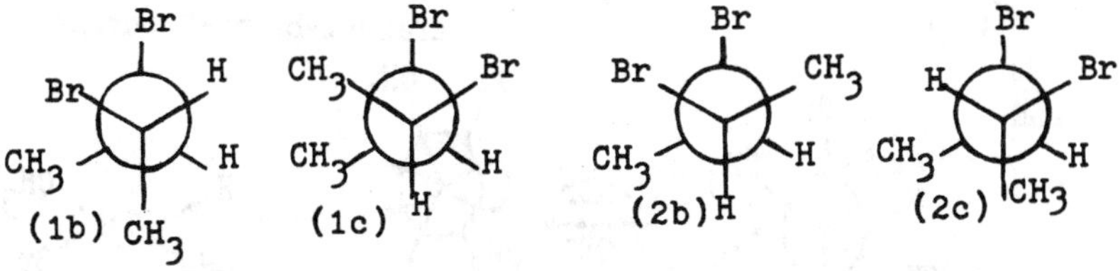

We are now asked to show the structures of the alkenes that could be formed from each by a trans *E*2 elimination of one mole of hydrogen bromide with hydroxide ion. This requires that in each conformation we look for a hydrogen trans to a bromine and then when making the *E*2 product, converting the projection to an almost planar figure to judge whether the alkene is cis or trans. For example, let us consider structure 1b:

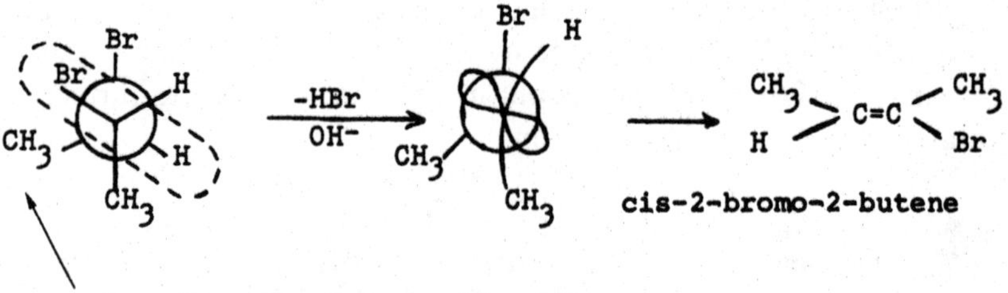

cis-2-bromo-2-butene

anti-conformation between H and Br

1c is done in the same manner:

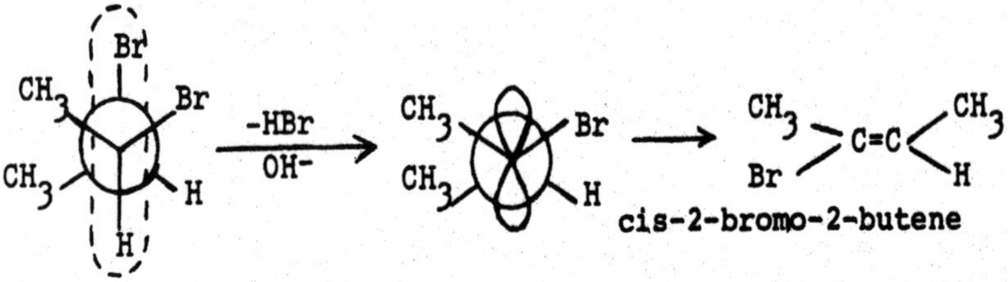

cis-2-bromo-2-butene

Conformations 2a, 2c, and 1a will not undergo trans *E*2 elimination because there are no hydrogens anti to a bromine. Conformation 2b has two possible trans elimination routes:

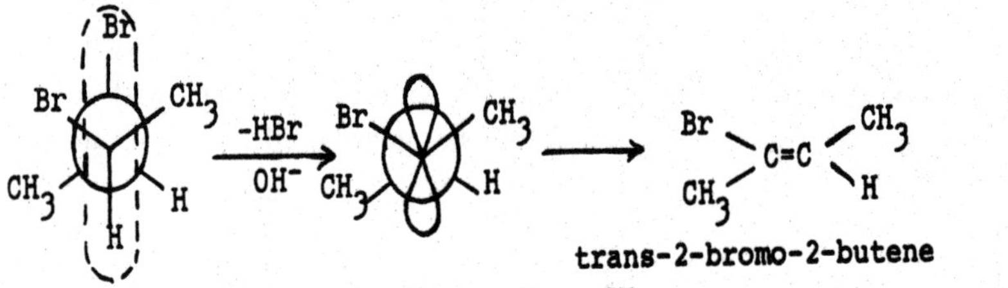

trans-2-bromo-2-butene

and

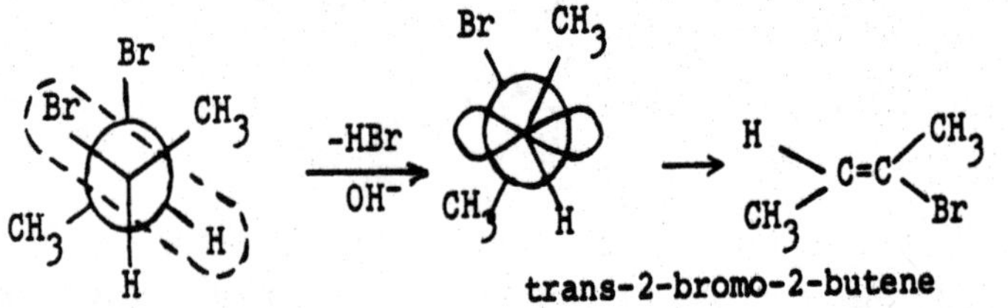

trans-2-bromo-2-butene

There is no significance to this since both form the same isomeric alkene, thus the two alkenes that form are trans- and cis-2-bromo-2-butene.

To see which isomer will eliminate more readily, we look at the Newman projections for each isomer. Only in the trans isomer is there still another anti H to Br conformation. In the cis isomer, H and Br are eclipsed and less stable than the anti conformation. Elimination via an eclipsed conformation is called syn elimination, same side, and it is generally less stable and slower than anti elimination. We conclude that the trans isomer (H and Br are anti to each other) will more readily eliminate further to form dimethylacetylene.

Q The reaction of (±) – 2,3-dibromobutane with ethoxide ion produces trans-2-bromo-2-butene while, under the same conditions, meso – 2,3-dibromobutane produces the corresponding cis isomer. With the aid of three-dimensional representations, determine whether this is an anti elimination. Explain how it is possible for an anti elimination to produce both cis and trans products.

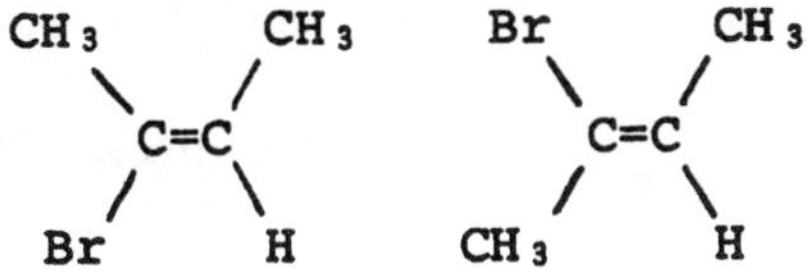

CH_3 CH_3 / Br H (cis) $\quad$ Br CH_3 / CH_3 H (trans) $\quad$ 2-bromo-2-butene

cis $\quad$ trans

A To solve this problem we should look at the structures of (±)–2,3-dibromobutane and meso-2,3-dibromo-butane in three-dimensional representations:

(and mirror image)

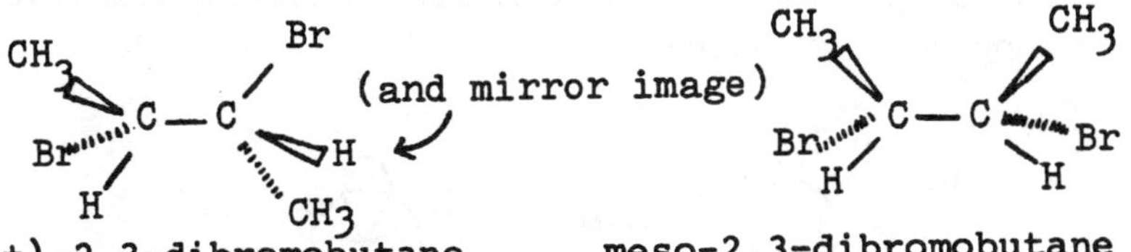

(±)-2,3-dibromobutane $\quad$ meso-2,3-dibromobutane

To make things as clear as possible, we will present the meso-2,3-dibromobutane as a Neuman projection:

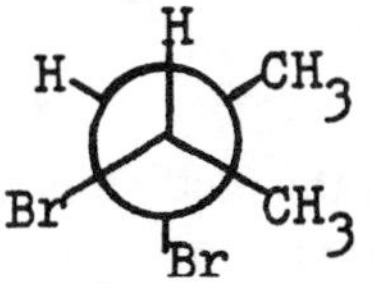

meso-2,3-dibromobutane

We have conveniently rotated this Newman projection so that it is ready to undergo an anti elimination (the H and Br are anti to each other). The point of this is that if we get a cis product we know that it does undergo an anti elimination, but if we get a trans product then this meso compound does not undergo an anti elimination. We now go through this elimination (*E*2):

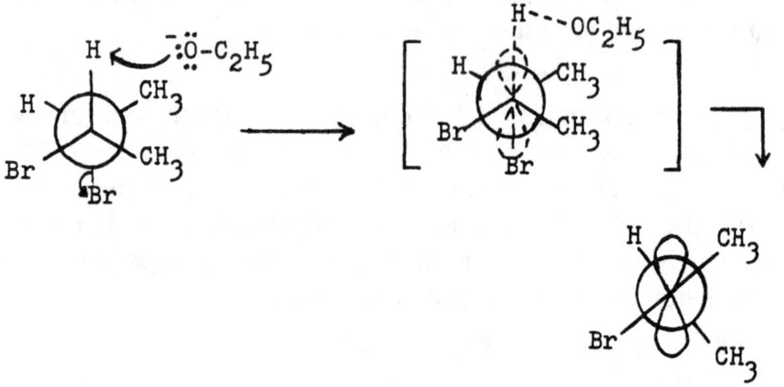

cis-2-bromo-2-butene

We can conclude from this representation that the elimination reaction of meso-2,3-dibromobutane to form cis-2-bromo-2-butene proceeds via an anti elimination.

To explain how it is possible for an anti elimination to produce both cis and trans products, let us examine the compound $CH_3CH_2CHBrCH_3$ by using the Newman projection:

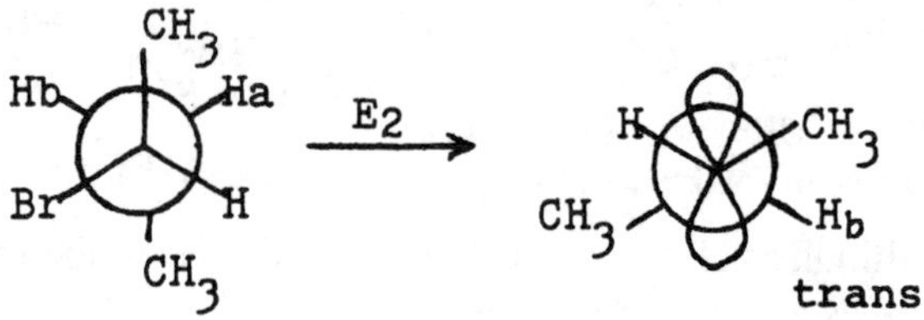

trans product

Figure 1

We labeled the two hydrogens of the methylene group to show two possible routes of an anti elimination, Figure 1 shows H_a being eliminated and a trans product being formed.

In Figure 2, by labeling the hydrogens to be eliminated, we see that there can be both cis and trans products.

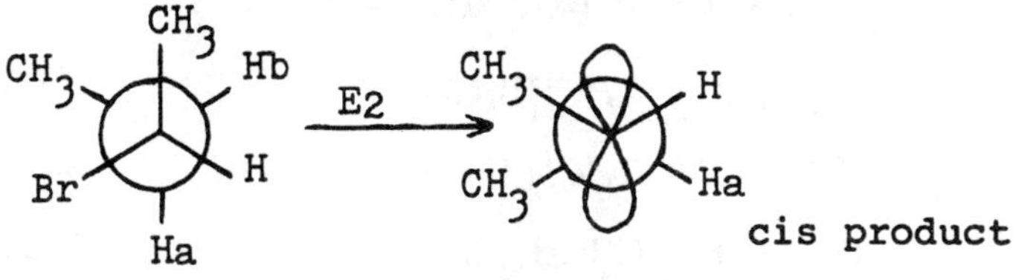

CH_3 CH_3 Hb Br H Ha $\xrightarrow{E2}$ CH_3 CH_3 H Ha

cis product

Figure 2

Quiz: Alkyl Halides

1. Which of the following reactions will yield t-pentyl bromide?

I. 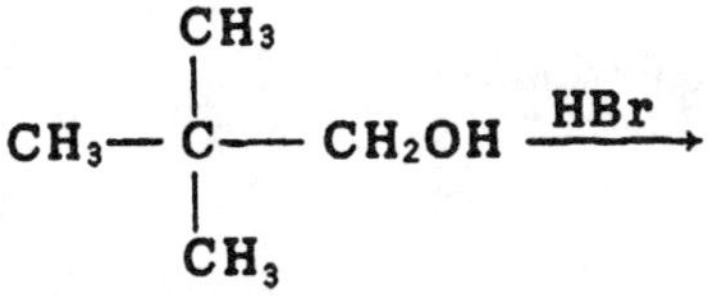 $CH_3-C(CH_3)_2-CH_2OH \xrightarrow{HBr}$

II. 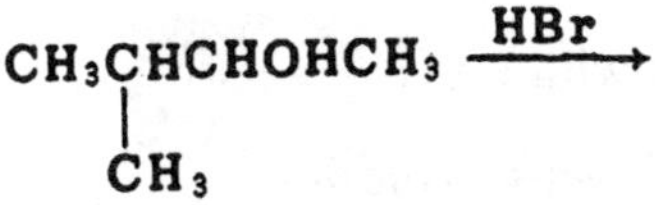 $CH_3CH(CH_3)CHOHCH_3 \xrightarrow{HBr}$

III. 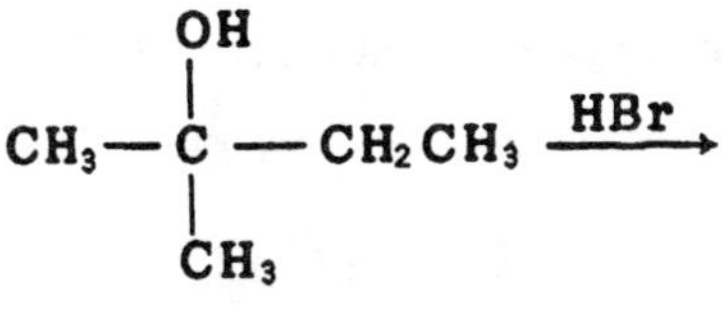 $CH_3-C(OH)(CH_3)-CH_2CH_3 \xrightarrow{HBr}$

(A) III only.

(B) II and III only.

(C) I and III only.

(D) All of the reactions.

(E) None of the reactions.

2. For the following reaction, product X is

$$(CH_3)_2CHCHOHCH_3 \xrightarrow{HBr} X$$

(A) $(CH_3)_2$ $CHCHBrCH_3$

(B) $(CH_3)_2$ $CHCH_2CH_2Br$

(C) $(CH_3)_2$ $CBrCH_2CH_3$

(D) $(CH_3)_2\underset{\displaystyle CH_3}{CH}CH{-}\underset{\displaystyle CH_3}{CH}CH(CH_3)$

(E) No reaction occurs.

3. Which one of the following reactions is not usually stereospecific?

(A) Free-radical substitution

(B) S_N1

(C) $E2$

(D) Hydrogenation with Pd/H_2

(E) Addition of halogens to olefins

4. Which one of the following is the most reactive in S_N2 reactions?

(A) $(CH_3)_3CCN$

(B) $(CH_3)_2CHCN$

(C) CH_3CH_2CN

(D) $(CH_3)_3CCl$

(E) CH_3CH_2Cl

5. Displacement reactions that proceed by the S_N2 mechanism are most successful with compounds that are

   (A) secondary halides with branches at C-2

   (B) neopentyl system.

   (C) primary compounds with no branches at the β-carbon.

   (D) tertiary compounds with no branches at the β-carbon.

   (E) primary halides with branches at the α-carbon.

6. Which one of the following is not true for the nucleophilic substitution referred to as the S_N2 reaction?

   (A) This reaction is of second-order kinetics.

   (B) There is complete stereochemical inversion.

   (C) There's absence of rearrangement.

   (D) Racemization is evident.

   (E) CH_3 $w > 1° > 2° > 3°$ (w refers to a substituent; 1°, 2°, and 3° refer to the carbon atoms).

7.

$$CH_3-\underset{\displaystyle CH_3}{\overset{|}{CH}}-\underset{\displaystyle Br}{CH}-CH_3 + Cl^- \xrightarrow{\text{acetone}}$$

(In the structure, the CH_3 branch is attached above the second carbon and Br below the third carbon.)

The significant product(s) of this reaction is (are):

I. $CH_3-CH(CH_3)-CH(Cl)-CH_3$

II.

$$CH_3-\overset{\overset{\large CH_3}{|}}{C}=CHCH_3$$

III.

$$CH_3-\overset{\overset{\large CH_3}{|}}{CH}-CH=CH_2$$

(A) I only.

(B) II and III only.

(C) I and II only.

(D) I, II, and III.

(E) None of these.

8. The reaction

D, H, CH_3CH_2, Br on C + KI $\xrightarrow{\text{acetone}}$

goes to completion and shows second-order kinetics. The major product is

(A) $CH_3CH=CH_2$.

(B) 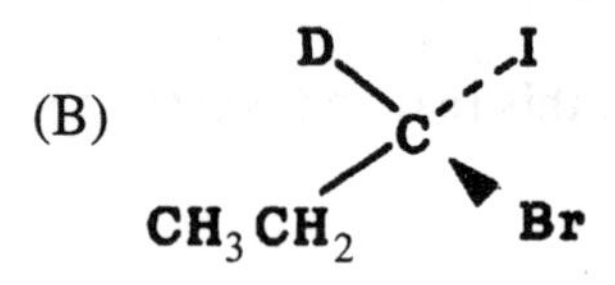

(C) [Wedge-dash structure: C bonded to I, H (dashed), Br (wedge), and CH_3CH_2]

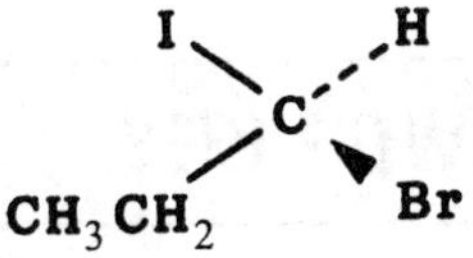

(D)

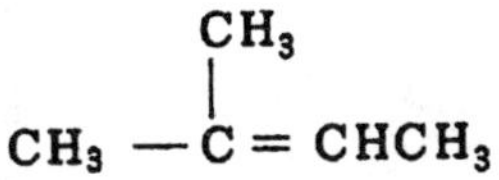

$$CH_3-\overset{\overset{CH_3}{|}}{C}=CHCH_3$$

(E)

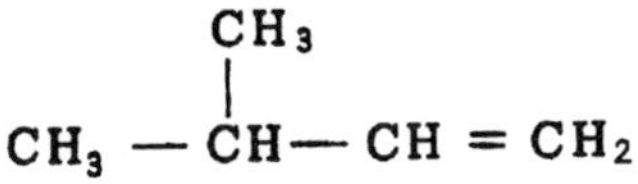

$$CH_3-\overset{\overset{CH_3}{|}}{CH}-CH=CH_2$$

9. The alkyl halide, $CH_3-CH_2-Br-CH_3$, is named

   (A) 1-bromobutane.
   (B) 2-bromobutane.
   (C) 1-bromopropane.
   (D) 2-bromopropane.
   (E) bromoethane.

10. Alkyl halides may be prepared from all of the following EXCEPT

   (A) fatty acids.
   (B) aromatic hydrocarbons.
   (C) alkenes.
   (D) alcohols.
   (E) methyl ketones.

ANSWER KEY

1. (D)
2. (C)
3. (A)
4. (E)
5. (C)
6. (D)
7. (D)
8. (E)
9. (D)
10. (B)

CHAPTER 6

Stereochemistry-Stereoisomerism

6.1 Optical Activity

Plane-polarized light is light whose vibrations take place in only one of the infinite planes through its line of propagation.

An optically active substance is one that rotates the plane of polarized light.

The rotation is measured and detected by an instrument called a polarimeter. The amount of rotation is dependent upon the number of molecules the light meets while passing through the polarimeter tube. Substances that rotate the plane of polarized light to the right are called dextrorotatory and are symbolized by *d* or +. Their mirror images that rotate light to the left are called levorotatory and are symbolized by ℓ or −.

Specific rotation is the number of degrees of rotation observed if a 1-decimeter tube is used, and the compound being examined is present to the extent of 1 g/cc.

$$[\alpha] = \frac{\alpha}{1 \times d}$$

$$\text{specific rotation} = \frac{\text{observed rotation (degrees)}}{\text{length (dm)} \times \text{g/cc}}$$

where *d* represents density for a pure liquid or concentration for a solution.

Problem Solving Examples:

Q Draw stereochemical formulas for all the possible stereoisomers of the following compounds. Label pairs of enantiomers and meso compounds. Tell which isomers, if separated from all other stereoisomers, will be optically active.

(a) 1,2-dibromopropane (b) 1,2,3,4-tetrabromobutane

A A knowledge of enantiomers, diastereomers, meso compounds, chirality, and optical activity will be necessary to solve this problem.

Stereoisomers may be defined as isomers that are different from each other only in the way the atoms are oriented in space. Nonsuperimposable mirror-image stereoisomers are referred to as enantiomers.

mirror

x, w, y, z | y, w, x, z

These two models are not superimposable. Although two substituent groups may coincide as they are twisted and turned, the other two do not. When molecules are not superimposable on their own mirror images, they are chiral. A compound whose molecules are chiral can exist as enantiomers; a compound whose molecules are without chirality (achiral) cannot exist as enantiomers. Enantiomers have identical physical and chemical properties, except for the direction of rotation of the plane of polarized light and the reaction with optically active reagents. Because enantiomers can rotate the plane of polarized light, they are called optically active substances. Stereoisomers that are not mirror images of each other are called diastereomers. A meso compound is defined as one whose molecules are superimposable on their mirror images, even though they contain chiral centers. Therefore, meso compounds contain chiral centers but nevertheless are not chiral. For example, 2,3-dichlorobutane would be a meso compound with an internal plane of symmetry (indicated by a dotted line):

```
      CH3          :          CH3
       |           :           |
  H ---+--- Cl     :     Cl ---+--- H
  - - -+- - - -    :     - - - + - - -
  H ---+--- Cl     :     Cl ---+--- H
       |           :           |
      CH3          :          CH3
```

superimposable

In solving this problem, we need to draw the compounds and look for possible configurations that would give rise to enantiomers, diastereomers, or meso compounds. One method of drawing the compound is to use a cross in which the intersection marks the location of the chiral carbon and the four groups attached to the carbon are at the ends of the cross.

(a) 1,2-dibromopropane ($CH_3CHBrCH_2Br$). This structure may be drawn as follows:

```
          H
          |
  CH3 ----+---- CH2Br
          |
          Br
```

Note what happens if we rewrite this as

```
           H
           |
  BrCH2 ---+--- CH3
           |
           Br
```

This is indeed the mirror image of the first structure. Try to super impose these two forms. The fact is that they cannot be made to superimpose.

```
         H                    :              H
         |                    :              |
CH3 -----+----- CH2Br         :    BrCH2 ----+----- CH3
         |                    :              |
         Br                   :              Br
                              :
                           mirror
```

Hence, for 1,2-dibromopropane there exist two enantiomers, which when separated will rotate the plane of polarized light to the same degree but in opposite directions.

(b) 1,2,3,4-tetrabromobutane ($BrCH_2CHBrCHBrCH_2Br$). Three stereochemical formulas may be drawn in this case. Two of the three represent enantiomers that would be optically active if separated, whereas the third is an optically inactive meso compound.

```
      CH2Br
        |
   H ---+--- Br
   -----+-------
   H ---+--- Br
        |
      CH2Br
```

This configuration represents the meso compound. A plane of symmetry is evident as indicated by the dotted line.

```
      CH2Br          :         CH2Br
        |            :           |
   H ---+--- Br      :     Br ---+--- H
        |            :           |
  Br ---+--- H       :      H ---+--- Br
        |            :           |
      CH2Br          :         CH2Br

                   mirror
```

These two mirror images cannot be superimposed. They are enantiomers.

Q The concentration of cholesterol dissolved in chloroform is 6.15 g per 100 ml of solution. (a) A portion of this solution in a 5-cm polarimeter tube causes an observed rotation of –1.2°. Calculate the specific rotation of cholesterol. (b) Predict the observed rotation if the same solution were placed in a 10-cm tube. (c) Predict the observed rotation if 10 ml of the solution were diluted to 20 ml and placed in a 5-cm tube.

A An optically active substance may be defined as one that rotates the plane of polarized light. This rotation is measured and detected by an instrument called the polarimeter. The amount of rotation observed will reflect the number of molecules the light encounters in passing through the polarimeter tube. The specific rotation may be defined as the number of degrees of rotation observed if a 1-decimeter tube is used, and the compound being examined is present to the extent of 1 g/cc. The following equation is employed:

$$[\alpha] = \frac{\alpha}{l \times d}, \quad \text{that is,}$$

$$\text{specific rotation} = \frac{\text{observed rotation (degrees)}}{\text{length (dm)} \times \text{g/cc}},$$

where g/cc represents density for a pure liquid or concentration for a solution. To calculate (a), (b), and (c), we can use this equation and the given information.

(a) The observed rotation, α, is –1.2° in a 5-cm or 0.5-dm polarimeter. Since the concentration of cholesterol dissolved in chloroform is 6.15 g per 100 ml of solution,

$$d = \frac{6.15}{100} \frac{g}{ml}$$

Substituting:

$$[\alpha] = \frac{\alpha}{l \times d} = \frac{-1.2^\circ}{0.5 \times \frac{6.15}{100}} = -39.0^\circ$$

(b) Here, the same situation exists except that the tube is 10 cm or 1 dm, that is, it is twice as long. A doubled observed rotation should be anticipated, that is, $-\alpha$ should now be –2.4°. This is confirmed by the following calculation:

$$-39.0 = \frac{\alpha}{1.0 \times \frac{6.15}{100}}; \qquad \alpha = -2.4^\circ.$$

(c) Here, the concentration has been halved by diluting from 10 ml to 20 ml. Hence, it can be expected that the observed rotation should be halved from part (a) since the number of molecules is halved. In other words, α should be –0.6°. This is indeed found to be the case by calculation:

$$-39.0 = \frac{\alpha}{0.5 \times \frac{6.15}{200}}; \qquad \alpha = -0.6°.$$

6.2 Chirality

A chiral center (C*) is a carbon atom with four different groups attached to it.

The following are some examples of chiral carbons.

$C_2H_5-C^*H(CH_3)-CH_2OH$	$CH_3-C^*H(OH)-COOH$	$C_2H_5-C^*H(Cl)-CH_3$	$C_6H_5-C^*H(D)-CH_3$
2-Methyl-1-butanol	Lactic acid	sec-Butyl chloride	α-Deuterioethylbenzene

Chiral molecules are not superimposable on their mirror images. Not all molecules that contain a chiral center are chiral, and not all chiral molecules contain a chiral center. Chiral molecules do not have a plane of symmetry.

The maximum number of stereoisomers that are possible for a compound with *n* chiral centers is given by 2^n.

Problem Solving Examples:

(a) Neglecting stereoisomers for the moment, draw all isomers of the formula C_3H_6DCl. (b) Decide which of these are chiral.

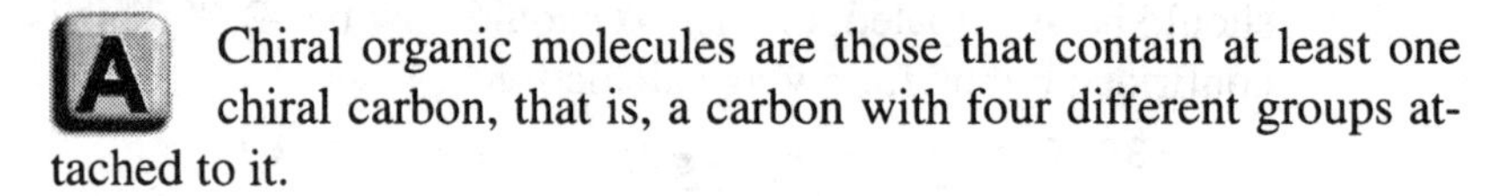

Chiral organic molecules are those that contain at least one chiral carbon, that is, a carbon with four different groups attached to it.

(a) The isomers of C_3H_6DCl are:

```
     H   H   Cl
     |   |   |
  H-C1 -C2 -C3 -D
     |   |   |
     H   H   H

       (A)
```

```
     H   Cl  D
     |   |   |
  H-C1 -C2 -C3 -H
     |   |   |
     H   H   H

       (B)
```

```
     Cl  H   D
     |   |   |
  H-C1 -C2 -C3 -H
     |   |   |
     H   H   H

       (C)
```

```
     H   Cl  H
     |   |   |
  H-C1 -C2 -C3 -H
     |   |   |
     H   D   H

       (D)
```

```
     Cl  D   H
     |   |   |
  H-C1 -C2 -C3 -H
     |   |   |
     H   H   H

       (E)
```

(b) To tell whether a molecule is chiral, one must locate the chiral carbon, if it exists. Thus, in structure (A), C3 is a chiral carbon and this isomer is chiral. In structure (B), C2 is a chiral carbon and this isomer is chiral. Isomers (C) and (D) contain no chiral carbons and are thus not chiral. Isomer (E) is a chiral molecule because C2 is a chiral carbon.

Which of the following formulas are chiral?

(a) 1-chloropentane

(b) 2-chloropentane

(c) 3-chloropentane

(d) 1-chloro-2-methylpentane

(e) 2-chloro-2-methylpentane

(f) 3-chloro-2-methylpentane

(g) 4-chloro-2-methylpentane

(h) 1-chloro-2-bromobutane

A To answer this question, one must first determine if a particular molecule or compound contains at least one carbon atom with four different groups attached to it. If it does, then the molecule is chiral.

(a) 1-chloropentane

```
     Cl  H H H H
     |   | | | |
  H-C---C-C-C-C-H
     |   | | | |
     H   H H H H
```

There is no chiral carbon in this molecule since no carbon contains four different substituent groups. The molecule is therefore not chiral.

(b) 2-chloropentane

```
   H   Cl  H   H   H
   |   |   |   |   |
H-C¹-C²-C³-C⁴-C⁵-H
   |   |   |   |   |
   H   H   H   H   H
```

C2 is a chiral carbon since it has four different groups attached to it: H_3C–, Cl–, H–, and $CH_3CH_2CH_2$. It can better be represented as:

```
      Cl
      |
CH3-C*-C3H7
      |
      H
```

(c) 3-chloropentane

```
  H H Cl H H
  | | |  | |
H-C-C-C--C-C-H
  | | |  | |
  H H H  H H
```

This molecule contains no chiral center since no carbon atom contains four different groups.

(d) 1-chloro-2-methylpentane

```
   Cl  H   H   H   H
   |   |   |   |   |
H-C¹-C²-C³-C⁴-C⁵-H
   |   |   |   |   |
   H   |   H   H   H
     H-C-H
       |
       H
```

C2 is a chiral carbon; thus this molecule is chiral. It can be better represented as:

```
            H
            |
ClCH2 ----C*----C3H7
            |
           CH3
```

(e) 2-chloro-2-methylpentane

```
    H    Cl   H    H    H
    |    |    |    |    |
  H-C ── C ── C ── C ── C-H
    |    |    |    |    |
    H    |    H    H    H
       H-C-H
         |
         H
```

This molecule contains no chiral center and is not chiral.

(f) 3-chloro-2-methylpentane

```
    H   H   Cl  H   H
    |   |   |   |   |
  H-C1 -C2 -C3 -C4 -C5 -H
    |   |   |   |   |
    H   |   H   H   H
      H-C-H
        |
        H
```

C3 is a chiral carbon and thus the molecule is chiral. It can be better represented as:

```
           Cl
           |
  C3H7 ── C* ── C2H5
           |
           H
```

(g) 4-chloro-2-methylpentane

```
    H   H   H   Cl  H
    |   |   |   |   |
  H-C1 -C2 -C3 -C4 -C5 -H
    |   |   |   |   |
    H   |   H   H   H
      H-C-H
        |
        H
```

C4 is a chiral carbon, and thus the molecule is chiral. It can be better represented as:

```
           Cl
           |
  C4H9 ── C* ── CH3
           |
           H
```

(h) 1-chloro-2-bromobutane

```
     Cl  H   H   H
     |   |   |   |
   H-C¹-C²-C³-C⁴-H
     |   |   |   |
     H   Br  H   H
```

C2 is the chiral carbon, and thus this molecule is chiral. It can be better represented as:

```
            H
            |
  ClCH₂——C*——C₂H₅
            |
            Br
```

Thus, only molecules (b), (d), (f), (g), and (h) contain chiral carbons and are thus chiral.

6.3 Enantiomerism

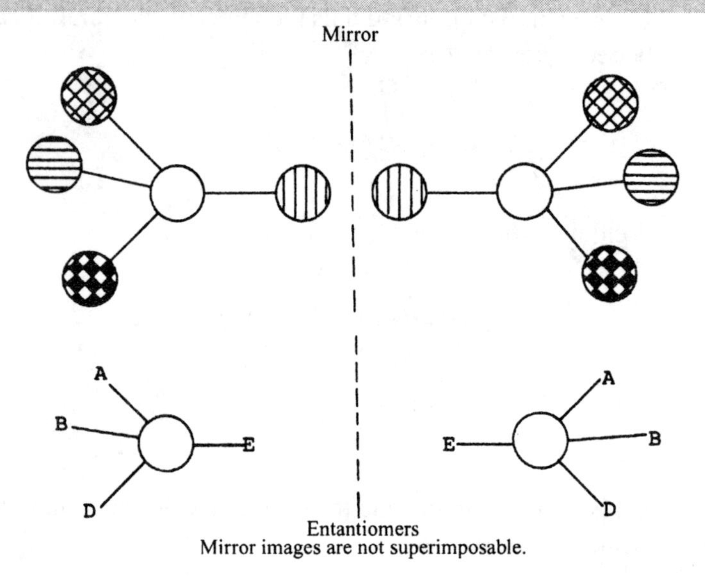

Entantiomers
Mirror images are not superimposable.

Enantiomers are isomers that are mirror images of each other, and they belong to the general class called stereoisomers. Stereoisomers are isomers that are different from each other only in the way the atoms are oriented in space.

Chiral molecules can exist as enantiomers, and achiral (without chirality) molecules cannot exist as enantiomers.

Enantiomers have identical physical properties except for the direction in which they rotate plane-polarized light. Enantiomers have identical chemical properties except toward optically active reagents.

A pair of an enantiotopic groups contains two identical substituents. When one of the substituents is replaced with another substituent an enantiomeric pair results.

Problem Solving Examples:

Identify one pair of enantiotopic groups in each of the following:

(a) CH_2ClBr

(b) $CHCl\ (CH_3)_2$

(c) $CHCl\ (CH_2CH_3)_2$

(d) $ClCH_2CH_2Br$

(e) $ClCH_2CH_2Cl$

(a) Replacement of either of the two methylene protons would give a pair of enantiomers.

H_a (top), Br—C—Cl, H_b (bottom) $\xrightarrow{R}$ [R (top), Br—C—Cl, H_b (bottom) ⋮ R (top), Cl—C—Br, H_a (bottom)]

H_a replaced $\qquad$ H_b replaced

Such pairs of protons (H_a and H_b) are called enantiotopic protons.

(b) CH_{3a} (top), H—C—Cl, CH_{3b} (bottom) $\xrightarrow{R}$ [R (top), H—C—Cl, CH_{3b} (bottom) ⋮ R (top), Cl—C—H, CH_{3a} (bottom)]

The two methyl groups are enantiotopic groups. So we see that not only protons (H) may be enantiotopic, but entire groups of atoms can be enantiotopic.

(c) Fischer projection: top C_2H_{5a}, left H, right Cl, bottom C_2H_{5b} $\xrightarrow{R}$ [top R, left H, right Cl, bottom C_2H_{5b} : top R, left Cl, right H, bottom C_2H_{5a}]

Replacement of either ethyl group gives a pair of enantiomers. Thus, the two ethyl groups are enantiotopic.

(d) Fischer projection: top H_a, left Cl, right CH_2Br, bottom H_b $\xrightarrow{R}$ [top R, left Cl, right CH_2Br, bottom H_b : top R, left $BrCH_2$, right Cl, bottom H_a]

and

Fischer projection: top H_c, left $ClCH_2$, right Br, bottom H_d $\xrightarrow{R}$ [top R, left $ClCH_2$, right Br, bottom H_d : top R, left Br, right CH_2Cl, bottom H_c]

Thus, there are two pairs of enantiotopic protons: H_a and H_b, and H_c and H_d.

(e) $ClCH_2CH_2Cl$. In the same way as in problem (d), we can see that the hydrogens of either methylene group are enantiotopic hydrogens.

Q Draw the structures of all the different staggered conformations possible for (+)-tartaric acid. Are any of these identical with their mirror images? How many optically active forms of tartaric acid could there be altogether if rotation were not possible about the $C_2 - C_3$ bond and only the staggered conformations were allowed?

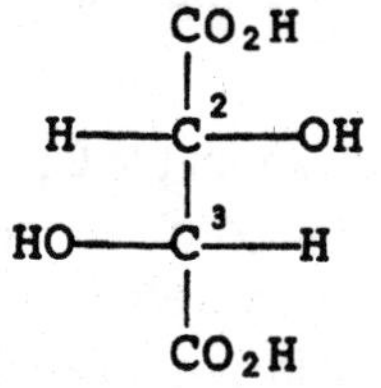

$$
\begin{array}{c}
CO_2H \\
| \\
H-\overset{2}{C}-OH \\
| \\
HO-\overset{3}{C}-H \\
| \\
CO_2H
\end{array}
$$

(+)-tartaric acid

Different conformations of a molecule result from rotation about single bonds. The bonding between atoms does not change, so different conformational isomers have the same configuration about their chiral centers.

Three different staggered conformations of (+)-tartaric acid result from rotation about the C_2–C_3 bond. These (along with their mirror images) are:

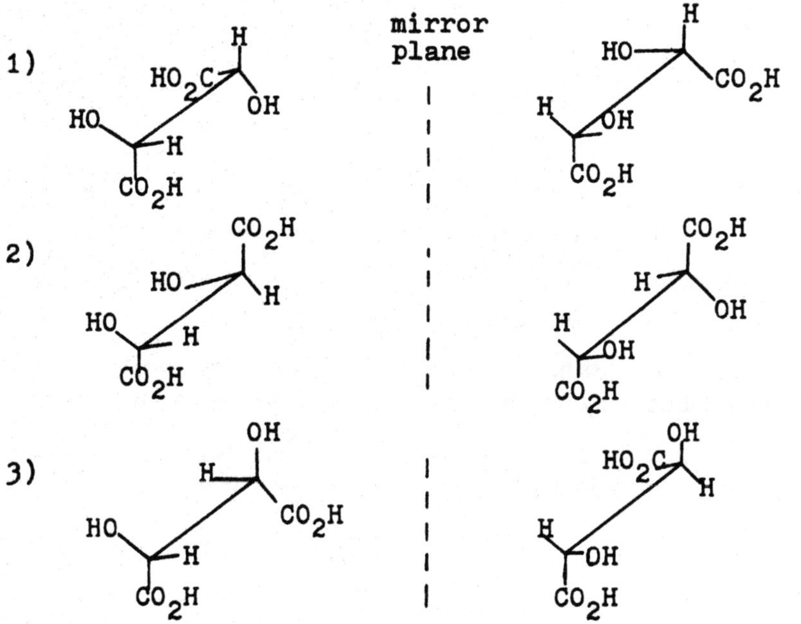

mirror plane

1)

2)

3)

(+)-tartaric acid

(–)-tartaric acid

None of the staggered conformations of (+)-tartaric acid has superimposable mirror images.

If rotation were not possible about the $C_2 - C_3$ bond, then any molecule of tartaric acid would be "locked" in one conformation. Assuming that only staggered conformers are allowed, there would be three enantiomeric pairs of tartaric acid possible (or a total of six isomers that are staggered).

If rotation about the $C_2 - C_3$ bond were allowed to occur, interconversion between the conformers would render the three forms indistinguishable from each other.

6.4 Diastereomers

Stereoisomers that are not mirror images of each other are called diastereomers.

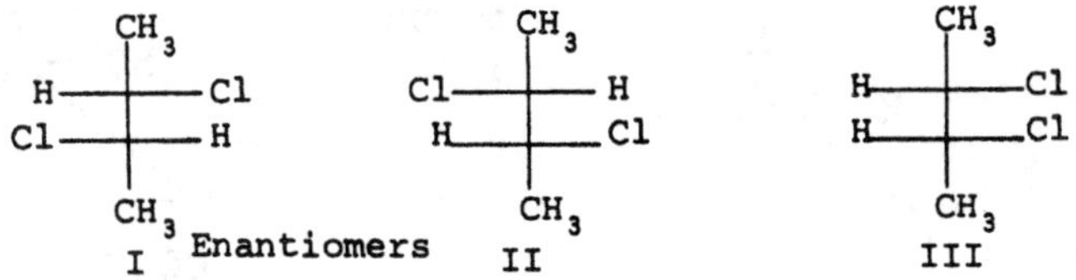

```
        CH3                      CH3                       CH3
         |                        |                         |
   H ----+---- Cl         Cl -----+----- H          H ------+------ Cl
  Cl ----+---- H           H -----+----- Cl         H ------+------ Cl
         |                        |                         |
        CH3                      CH3                       CH3
         I     Enantiomers        II                        III
```

III is a diastereomer of I and II where indicated.

Diastereomers have similar (not identical) chemical properties, since they are members of the same family.

Diastereomers have different physical properties and differ in specific rotation; they may be levorotatory, dextrorotatory, or inactive.

The particular class of diastereomers that results from restricted rotation about a double bond are called geometric isomers.

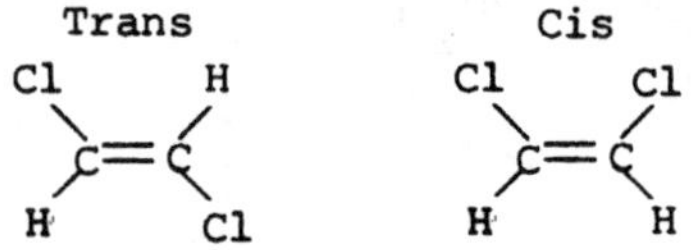

```
        Trans                    Cis
   Cl          H           Cl          Cl
     \        /              \        /
      C  ==  C                C  ==  C
     /        \              /        \
   H           Cl          H           H
```

1,2-Dichloroethane

I and II are geometric isomers.

Geometric isomers have different physical properties and can be separated from each other.

A meso compound is one whose molecules are superimposable on their mirror images even though they contain chiral centers. A meso compound is optically inactive.

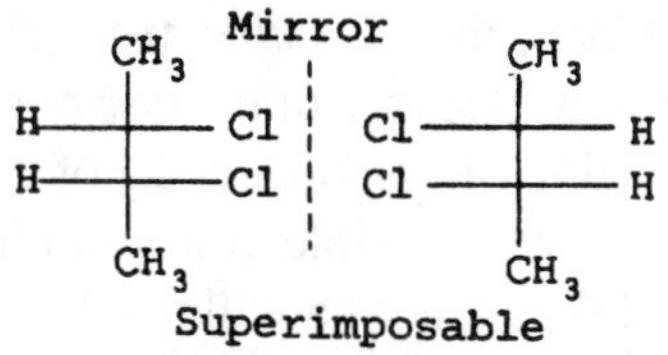

```
                Mirror
       CH3        |          CH3
        |         |           |
  H ----+---- Cl  |  Cl ------+------ H
  H ----+---- Cl  |  Cl ------+------ H
        |         |           |
       CH3                   CH3
             Superimposable
```

We can often recognize a meso compound on sight by the fact that half the molecule is the mirror image of the other half.

Problem Solving Examples:

Q Write structures showing the configuration of each of the possible products to be expected from the following reactions. Which diastereomer would you expect to be formed preferentially, assuming that the substituents C≈crease in size in the order $C_6H_5 > C_2H_5 > CH_3 > OH > H$?

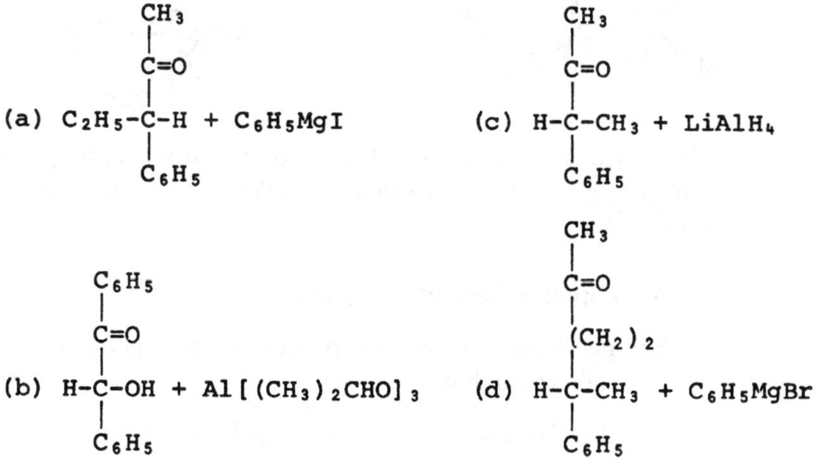

```
          CH3                                   CH3
          |                                     |
          C=O                                   C=O
          |                                     |
(a) C2H5-C-H + C6H5MgI                (c) H-C-CH3 + LiAlH4
          |                                     |
          C6H5                                  C6H5

                                                CH3
                                                |
     C6H5                                       C=O
     |                                          |
     C=O                                        (CH2)2
     |                                          |
(b) H-C-OH + Al[(CH3)2CHO]3           (d) H-C-CH3 + C6H5MgBr
     |                                          |
     C6H5                                       C6H5
```

A An understanding of the mechanism of the four reactions is necessary in order to be able to predict the predominating product of each reaction.

All four reductions involve attack of the carbonyl group. Since the atoms of the carbonyl group lie in a plane, attack can occur from either of the two sides of the plane, that is, either from above or below the plane and perpendicular to it. Because of the two possible directions of attack, the resultant tetrahedral carbon (originally of the carbonyl group), if chiral, will have one of two possible configurations (R or S). The predominating product of the reduction will result from the more sterically favored and thus more frequent attack.

(a)

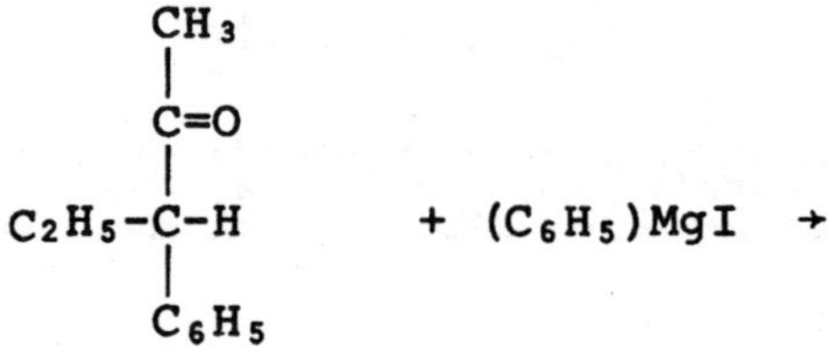

First, it is helpful to picture the compound in a three-dimensional manner. (Use models if necessary.)

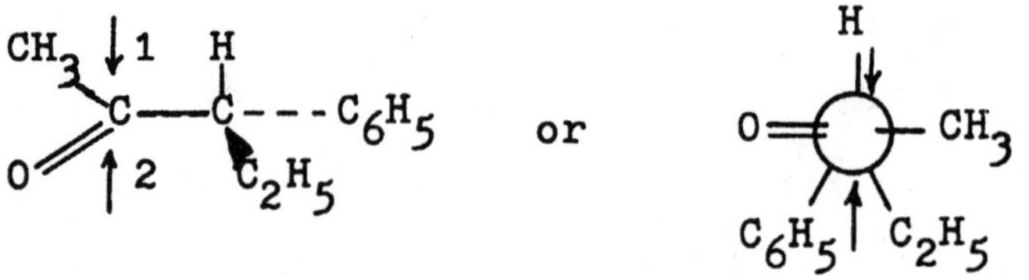

The two possible directions of attack of the carbonyl group are indicated by arrows so we have the following possible reaction mechanisms:

(A) compare their melting points.

(B) place one of them into solution and pass polarized light through it.

(C) dissolve each one (individually) in a particular solvent and then compare their respective solubili.

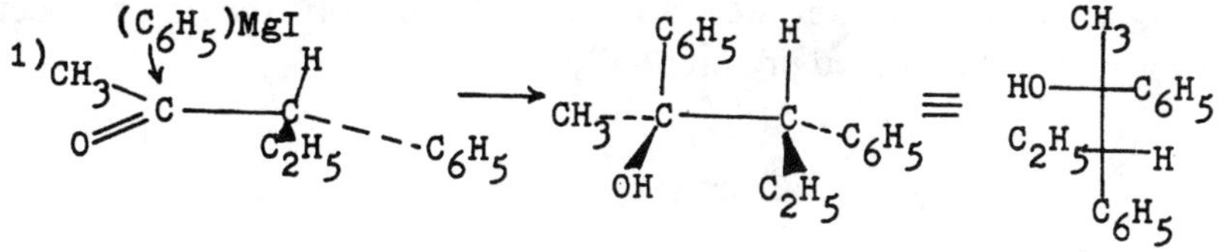

A

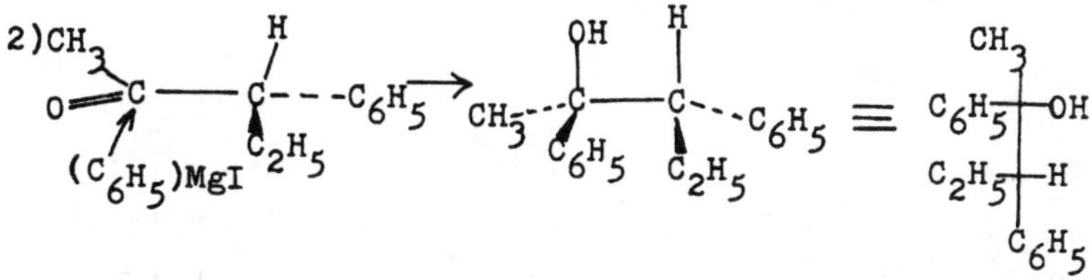

B

As we can see, attack 2 will have to overcome more steric hindrance (due to $-C_2H_5$ and $-C_6H_5$ groups) than attack 1. Therefore, the A product will predominate over the B product.

(b)

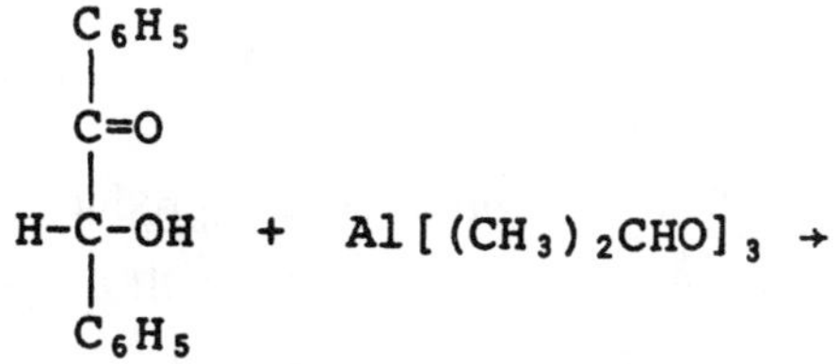

$$C_6H_5-C(=O)-CH(OH)-C_6H_5 + Al[(CH_3)_2CHO]_3 \rightarrow$$

Again, the two possible directions of attack will lead to two products.

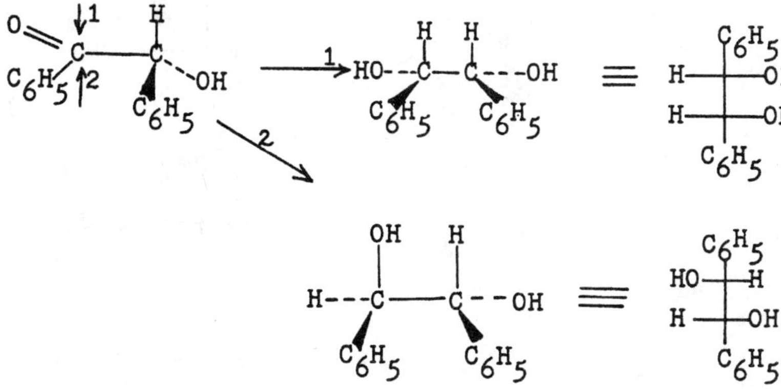

(1)

(2)

Attack 1 is more sterically favored; therefore, the first product will be formed preferentially.

(c)

```
        CH3
        |
        C=O
        |
      H-C-CH3      + LiAlH4 →
        |
        C6H5
```

```
      H
  O   ↓1      H                 1              H     H                        CH3
   \\ |       |                 →         HO---C-----C--CH3      ≡      H-----|-----OH
     C--------C---CH3                          |     |                  H-----|-----CH3
    /  ↑2      \                               CH3   C6H5                     |
 CH3            C6H5                                                          C6H5     (1)
                                2
                                ↘              OH    H                        CH3
                                         H----C-----C---CH3        ≡    HO----|-----H
                                              |     |                   H-----|-----CH3
                                              CH3   C6H5                      |
                                                                              C6H5     (2)
```

Product 1 will be formed in preference to product 2.

(d)

```
                                                              CH3
                                                              |
                                                       C6H5---|-------OH
      CH3                                   1                 |
      |                                   ↗                  (CH2)2
      C=O                                                     |
      |          + C6H5MgBr                                   |
      CH2                                              H------|-------CH3
      |                                                       |
      CH2                                                     C6H5           (1)
      |                                     2
    H-C-CH3                                 ↘                 CH3
      |                                                       |
      C6H5                                             HO-----|-------C6H5
                                                              |
                                                             (CH2)2
                                                              |
                                                       H------|-------CH3
                                                              |
                                                              C6H5           (2)
```

Due to the presence of the two – CH_2 groups, which increases the distance between the carbonyl group and the bulky – CH_3 and – C_6H_5 substituents, there is no significant difference in the steric environment of the carboxyl group in formation of either product. Thus, 1 and 2 will be formed in approximately equal amounts.

In the following structures, indicate if H_a and H_b are identical, enantiotopic, or diastereotopic.

(a) CH_3; H_a—OH; HO—H_b; CH_3

(b) CH_3, H_a; CH_3, H_b

(c) CH_3; H_b H_a

(d) CH_3; H_b—OH; HO—H; H_a—OH; CH_3

(e) H CH_3; H; Br; H; H_b; H_a

(f) H; H_b—C—S—H; H_a

(a) If we replace either H_a or H_b, we get identical structures; therefore, we can call these two hydrogens identical.

(b) Here H_a and H_b are diastereotopic protons. Note that one hydrogen (H_b) is axial while the other (H_a) is equatorial. Since the environments of these two protons are not mirror images and lack a chiral center, replacing either of the two Hs will yield a pair of diastereomers:

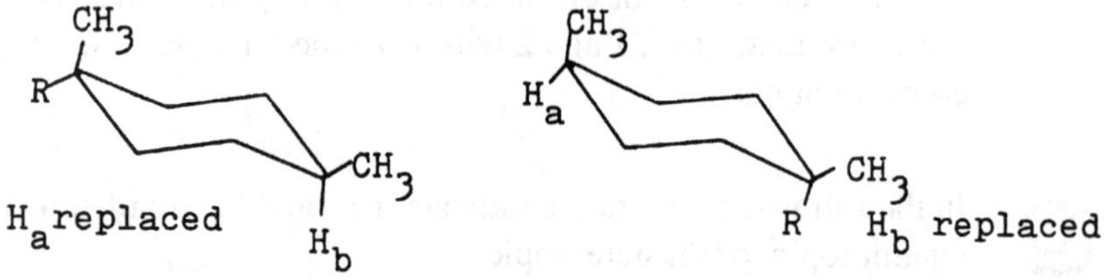

CH_3 R CH_3 H_b — H_a replaced

CH_3 H_a CH_3 R — H_b replaced

These two protons can be identical if we allow the ring to undergo inversion.

(c) H_a and H_b are enantiotopic protons in this structure. Replacement of either of the two protons results in a pair of enantiomers:

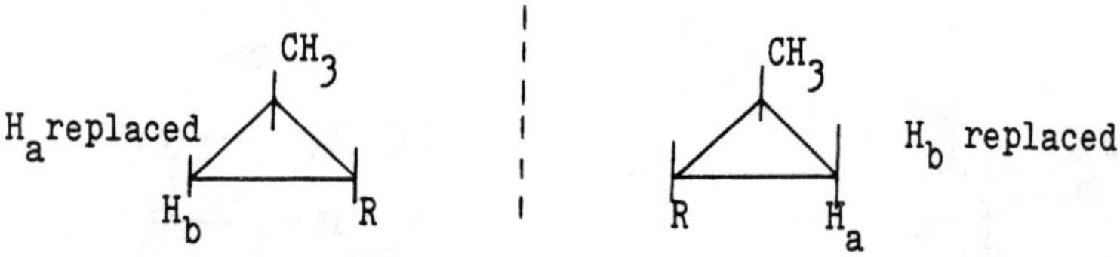

CH_3 H_b R — H_a replaced

CH_3 R H_a — H_b replaced

Note that if the Hs were positioned on the same side of the ring as – CH_3, the protons would be identical.

(d) Notice that both the carbons to which H_a and H_b are attached are chiral. Replacement of either H_a or H_b gives us a pair of enantiomers; hence, the hydrogens are enantiotopic.

(e) H_a and H_b are enantiotopic. The reason is the same as that given in (d).

(f) The carbon to which H_a and H_b are attached is achiral. Replacement of either H_a or H_b gives us two identical structures and so the protons are identical.

6.5 Racemization

Racemization is the process that leads to the formation of a racemic modification, a mixture of equal parts of enantiomers.

A racemic modification is optically inactive because the rotating influence of one enantiomer just cancels that of the other.

The separation of a racemic modification into its component enantiomers is called resolution.

There are three methods commonly used to carry out resolutions: mechanical separation of asymmetric crystals, resolution by formation of diastereomers, and resolution by reaction with optically active reagents. The last method mentioned is the most widely used.

Problem Solving Example:

Q Compound A racemizes readily on heating to 100°, but the rate is not affected by chloride ion and is the same in chloroform and acetic acid. Racemization in deuteroacetic acid (CH_3CO_2D) gives only undeuterated racemic A. Devise a mechanism for the reaction in accord with all the experimental facts.

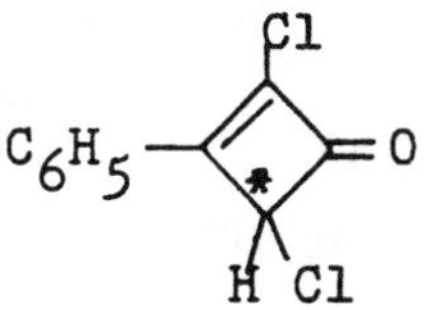

Cl

C_6H_5 =O

H Cl

(A)

A Racemization is the process that leads to formation of a racemic modification, a mixture of equal parts of enantiomers.

In compound A, there is only one asymmetric carbon (*). Chloride ion could promote racemization by an S_N2 = type of reaction.

Ring Ring

Cl^- C Cl ⇌ Cl—C---H + Cl^-

H Ring Ring

Attack by Cl^- inverts the configuration at the chiral carbon. Since the reaction is reversible, a racemic mixture will form as equilibrium is established. We are told, however, that chloride ion does not affect the rate of racemization. This is due likely to hindrance of chloride ion attack by the ring structure. Therefore, racemization probably does not proceed via the S_N2 mechanism.

Chloroform and acetic acid are both acidic solvents and may cause racemization by way of enol-keto tautomerism:

C_6H_5 H Cl =O H^+ ⇌ C_6H_5 O–H H^+ Cl ⇌ C_6H_5 =O Cl H

keto　　enol　　keto

If the racemization indeed proceeded via enol-keto interconversion, deuteroacetic acid could give deuterated racemic A in the following manner:

C_6H_5 =O H Cl ⇌ C_6H_5 –O⁻ Cl ⇌ D^+ → C_6H_5 =O D Cl ; C_6H_5 =O Cl D

racemic modification

However, reaction with CH_3CO_2D does not result in any deuterated A, so we can conclude that racemization does not occur via enolization.

The racemization is therefore most reasonably explained by a simple thermal reaction at 100°. Heat could cause reversible ring opening in the following manner:

C_6H_5 =O H Cl ⇌(Δ) C_6H_5–C CH C=O CHCl ⇌ C_6H_5 =O Cl H

Since the reaction is reversible, equilibrium will give the racemic mixture. Note that the reaction proceeds via electron exchange, and no substitution or addition of atoms occurs. Therefore, the reaction rate would not be affected by acids or chloride ions.

6.6 Specification of Configuration

The arrangement of atoms that characterizes a particular stereoisomer is called its configuration.

For specifying particular configurations, we follow two steps:

Step 1. A sequence of priority according to rules is assigned to the four atoms or groups of atoms attached to the chiral carbon.

Step 2. The molecule is oriented so that the group of lowest priority is directed away from us. The arrangement of the remaining groups is then observed. If our eye travels in a clockwise direction in going from the group of highest priority to the groups of lower priority, the configuration is specified by R. If eye movement is counterclockwise, the configuration is specified by S.

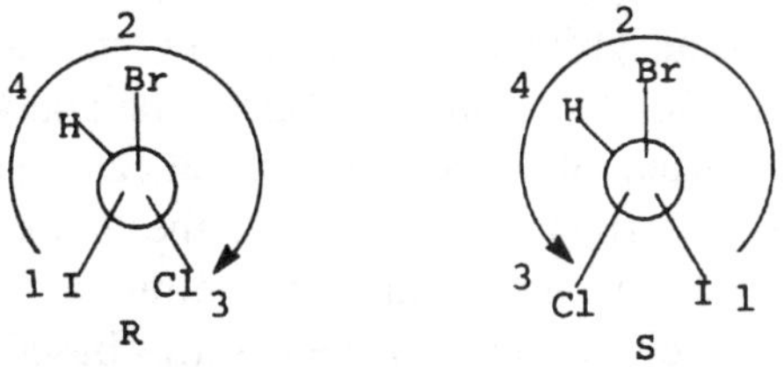

R	S
1 I, 2 Br, 3 Cl, 4 H	1 I, 2 Br, 3 Cl, 4 H

Sequence Rule 1. If the four atoms attached to the chiral center are all different, then priority depends on atomic number, with the atom of higher atomic number having priority. In the case of two atoms being isotopes of the same element, the atom of higher mass number has priority.

Sequence Rule 2. If priority cannot be decided from the first rule, then it will be determined by similar comparison of the atoms attached to it and so on.

Sequence Rule 3. Both atoms are considered to be duplicated or triplicated when there exists between them a double or triple bond.

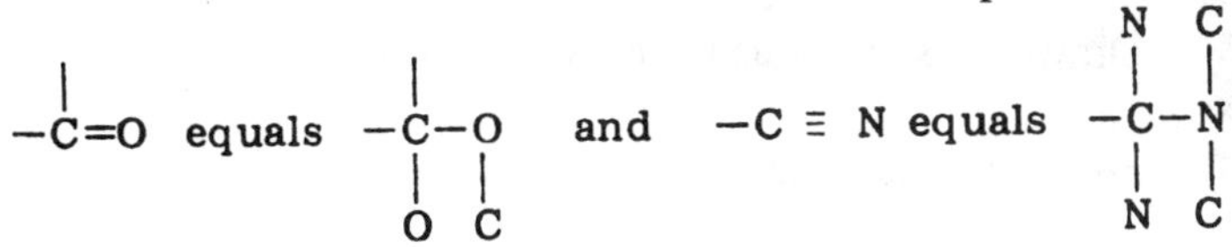

$-\overset{|}{C}=O$ equals $-\underset{O}{\underset{|}{\overset{|}{C}}}-\underset{C}{\underset{|}{O}}$ and $-C \equiv N$ equals $-\underset{N}{\underset{|}{\overset{N}{\overset{|}{C}}}}-\underset{C}{\underset{|}{\overset{C}{\overset{|}{N}}}}$

Configurations for compounds with more than one chiral center are specified by specifying the configuration for each of the centers and by using the number of the chiral center.

CH_3 / Cl—H / H—Cl / C_2H_5 (2R,3R)-2,3-Dichloropentane

CH_3 / H—Cl / H—Cl / C_2H_5 (2S,3R)-2,3-Dichloropentane

Fischer Convention

D or L isomers are used to represent the Fischer formula. D or L isomers are either positive or negative depending on the particular compound involved.

An isomer belongs to the D series if, when it is written with the aldehyde-containing group (or a related group such as COOH) attached to the chiral carbon at the top of the molecule and with the other carbon-containing group attached to the chiral carbon at the bottom, the OH group is on the right side. Its enantiomer belongs to the L series. When a compound has more than one chiral carbon atom, the compound belongs to either the D or the L series based on the configuration at the highest-numbered chiral carbon atom (farthest removed from the COH group).

H—C—OH ≡ H—OH (CH_2OH) D (+) -Glyceraldehyde

HO—H ≡ HO—C—H (CH_2OH) L (-) -Glyceraldehyde

Fischer Projection Formulas

Formulas

Problem Solving Examples:

Draw and specify as R or S the enantiomers of:

(a) 3-chloro-l-pentene

(b) 3-chloro-4-methyl-l-pentene

(c) $HOOCCH_2CHOHCOOH$, malic acid

(d) $C_6H_5CH\ (CH_3)\ NH_2$

A This problem deals with configuration, the arrangement of atoms that characterizes a particular stereoisomer. One method of assigning configuration, called the Cahn-Ingold-Prelog Method, is by use of the prefixes R and S. Two steps are involved in this process: In step 1, a set of sequence rules is followed. A sequence of priority is assigned to the four atoms or groups of atoms attached to the chiral center. (Recall, an atom to which four different groups are attached is a chiral center.) In step 2, the molecule is oriented so that the group of lowest priority is directed away from us. We observe, then, the arrangement of the remaining groups. If our eye travels in a clockwise direction in going from the group of highest priority to the groups of lower priority, the configuration is specified R. If eye movement is counterclockwise, the configuration is specified S.

At this point, the problem is how do we assign priority, that is, what are the sequence rules? For simplicity, we can note three sequence rules:

Sequence Rule 1: If the four atoms attached to the chiral center are all different, then priority depends on atomic number, with the atom of higher atomic number getting higher priority. In the case of two atoms being isotopes of the same element, the atom of higher mass number has the higher priority.

Sequence Rule 2: If sequence rule 1 fails to establish priority, then a similar comparison of the next atoms attached to the groups is made. In other words, if two atoms attached to the chiral center are the same, the atoms attached to each of these first atoms are compared.

Sequence Rule 3: Both atoms are considered to be duplicated or triplicated when there exists between them a double or triple bond. For example:

```
  H              H                          N C
  |              |                          | |
 -C=O  equals  -C-O  and  -C≡N  equals   -C-N
                 | |                        | |
                 O C                        N C
```

Now that we know how to specify configuration, we need to know a method that allows us to draw these enantiomers. One method is to draw a cross (+) and attach to the four ends the four groups that are bonded to the chiral center; the point where the lines cross denotes the location of the chiral center. For example, 2-chlorobutane (CH_3CHCl CH_2CH_3) may be represented by

```
      C2H5                 C2H5
       |                    |
   H---+---Cl    or    Cl---+---H
       |                    |
      CH3                  CH3
```

which are enantiomeric forms.

It is important to recognize the fact that in such representation the horizontal lines denote bonds coming toward us out of the plane of the paper, whereas the vertical lines represent bonds going away from us behind the plane of the paper.

With this information, it is now possible to draw and specify as R or S the following enantiomers (mirror-image isomers):

(a) 3-chloro-l-pentene ($CH_2 = CHCHClCH_2CH_3$)

```
          Cl                          Cl
          |                           |
   C2H5---+---CH=CH2       CH2=CH-----+---C2H5
          |                           |
          H                           H

          R                           S
```

Explanation: These enantiomers are drawn so that the chiral carbon of each structure is located at the intersection of the lines. The four groups attached to the chiral center are positioned at the four ends. To establish configuration, the priority of the groups must be determined. To do this recall that sequence rules must be followed. As mentioned in sequence rule 1, priority depends on atomic number. Chlorine has the highest atomic number, so it has the highest priority. Hydrogen, with the lowest atomic number, has the lowest priority. The two other atoms bonded to the chiral center are carbon atoms. But note that one carbon atom is involved in a double bond, so (in accordance with rule 3) it is duplicated and has the higher priority. Overall, then, we can rank in increasing priority the following groups of atoms:

$$\underrightarrow{\text{H, } C_2H_5\text{, } CH{=}CH_2\text{, Cl}}_{\text{increasing priority}}$$

At this point, recall that the group of lowest priority must be directed away from us. Note that the drawings given accomplish this. The hydrogen atom, the group of the lowest priority, is at the end of a vertical line. Consequently, it is directed away from us behind the plane of the paper. Finally, note that in the first configuration (the one to the left), our eyes move in clockwise (↻) direction in going from the group of highest priority (– Cl) to the groups of second (-CH = CH_2) and third priority (– C_2H_5). Hence, this configuration is considered to be R. In the other configuration (the one to the right), our eyes move counterclockwise (↺), so that S should be assigned to this configuration.

In (b) – (d), the same kind of reasoning is used to obtain the following:

(b)

```
              Cl                              Cl
              |                               |
(CH3)2CH------+------CH=CH2    CH2=CH---------+------CH(CH3)2
              |                               |
              H                               H

      ↻ ∴ R                           ↺ ∴ S
```

(c)

```
              H                               H
              |                               |
    HOOC------+------CH2COOH   HOOCCH2--------+------COOH
              |                               |
              OH                              OH

      ↻ ∴ R                           ↺ ∴ S
```

(d)

```
              H                               H
              |                               |
      C6H5----+------CH3           CH3--------+------C6H5
              |                               |
              NH2                             NH2

      ↻ ∴ R                           ↺ ∴ S
```

Draw and label the enantiomers of each of the following compounds as either R or S:

(a) 4-chloro-2-pentene (c) isopentylbenzene

(b) isobutyl alcohol (d) phenylalanine

(a) 4-chloro-2-pentene

```
    H  H  H  Cl  H
    |  |  |  |   |
  H-C--C=C--C*--C-H
    |        |   |
    H        H   H
```

Since there is one chiral center (C*), two (2^1) enantiomeric forms exist for this compound:

```
  H  H  H  Cl               :            Cl        H  H  H
  |  |  |  |                :            |         |  |  |
H-C--C=C---+--------CH3     :     CH3----+---------C=C--C-H
  |        |                :            |              |
  H        H                :            H              H

        I                   :                  II
```

According to the rules of sequence, the order of priority for the four groups is:

```
           H  H  H
           |  |  |
Cl  >    H-C--C=C   >   CH3   >   H
           |
           H
```

H, the atom of the lowest priority, is visualized as extending away from us behind the plane of the paper. Now follow the three remaining groups in decreasing priority. In doing this, I leads our eye into a counterclockwise movement, thus the chiral carbon of this structure is assigned an absolute configuration of S. II leads our eye into a clockwise movement, thus II has an absolute configuration of R. Notice that since these two structures are enantiomers, they are assigned opposite absolute configurations as expected.

(b) Isobutyl alcohol

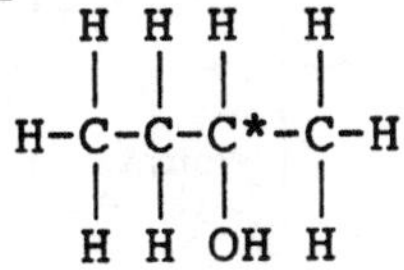

```
    H  H  H   H
    |  |  |   |
  H-C--C--C*--C-H
    |  |  |   |
    H  H  OH  H
```

C* is a chiral center. We can immediately see that two enantiomers are possible for this compound:

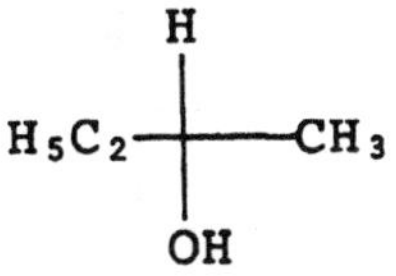

```
        H                          H
        |                          |
$H_5C_2$ —+— $CH_3$          $CH_3$ —+— $C_2H_5$
        |                          |
        OH                         OH

        I                          II
```

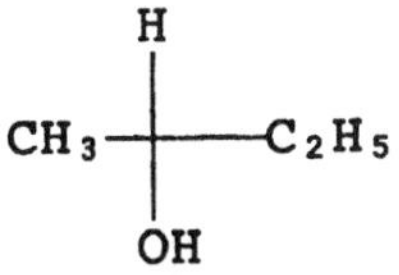

The order of priority is OH > C_2H_5 > CH_3 > H. With H visualized as projecting away from us, we can see that the remaining groups are arranged, according to decreasing priority, in a clockwise manner in I and in a counterclockwise manner in II. Thus, I has an absolute configuration of R while that of II is S.

Using the same reasoning, we can determine the configuration of the enantiomers of (c) and (d)

(c) Isopentylbenzene

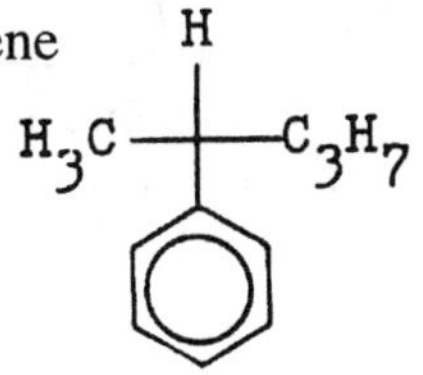

```
        H                          H
        |                          |
$H_3C$ —+— $C_3H_7$          $H_7C_3$ —+— $CH_3$
        |                          |
      (phenyl)                  (phenyl)

        I                          II
```

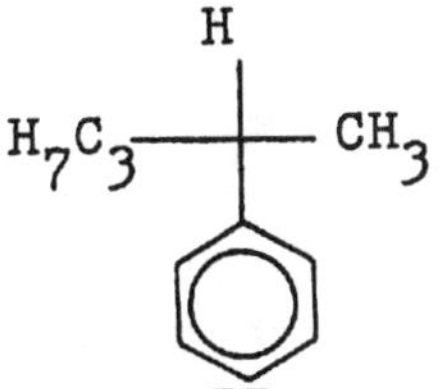

Priority	$C_6H_5 > C_3H_7 > CH_3 > H$	
Arrangement of substituents in decreasing priority	I counterclockwise	II clockwise
Configuration	S	R

(d) Phenylalanine

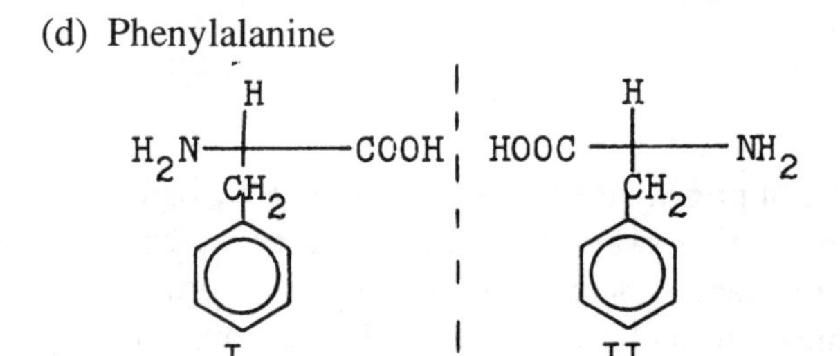

Priority	$NH_2 > COOH > CH_2C_6H_5 > H$	
	I	II
Arrangement	clockwise	counterclockwise
Configuration	R	S

Quiz: Stereochemistry-Stereoisomerism

1. An example of a dipole molecule is

(A) CH_4
(B) H_2
(C) H_2O
(D) NaCl
(E) O_2

2. In which one of the following compounds may cis-trans isomerism occur?

   (A) CH_4 (D) $C_2H_2(CH_3)_2$

   (B) C_6H_6 (E) C_2H_6

   (C) C_2H_4

3. Which of the following structures does not represent an optically active compound?

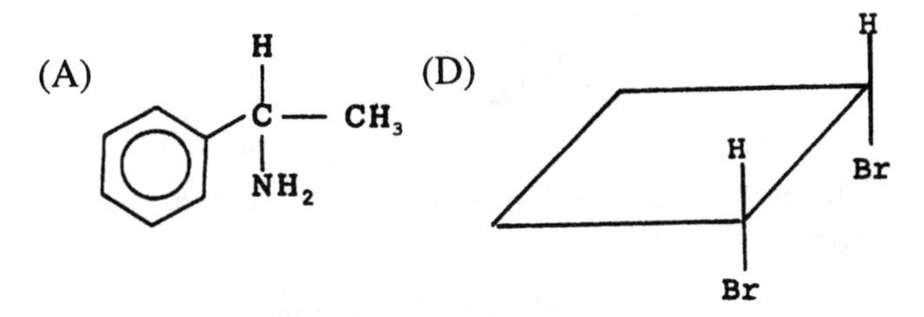

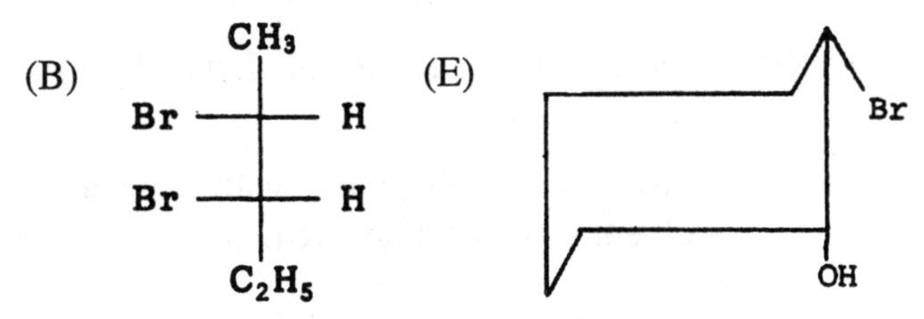

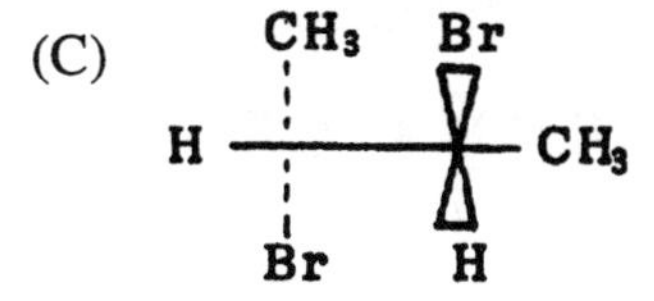

4. Which of the following is an R enantiomer of a chiral compound?

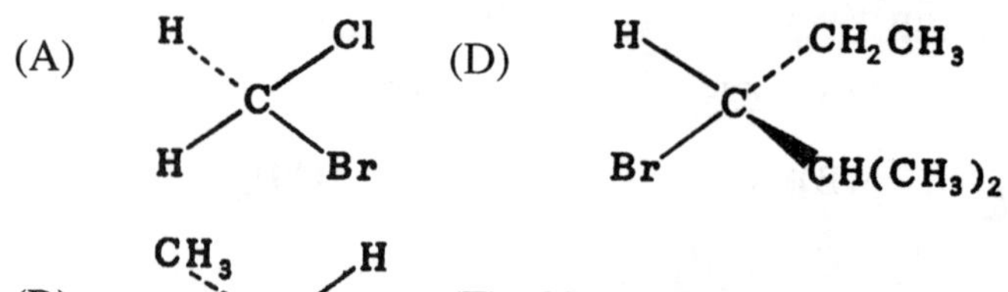

(B) CH_3, H, C, CH_3CH_2, Cl

(E) None of the above.

(C)

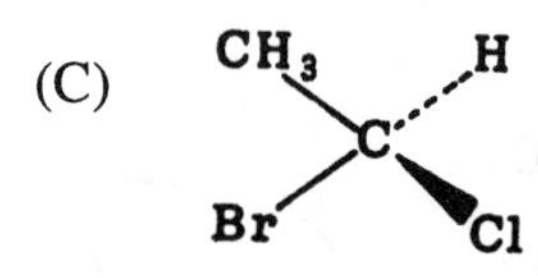

5. The best way to distinguish the presence of two enantiomers would be

(A) compare their melting points.

(B) place one of them into solution and pass polarized light through it.

(C) dissolve each one (individually) in a particular solvent and then compare their respective solubilities.

(D) examine them for color differences.

(E) compare their respective abilities.

6. Which of the following will not show cis-trans isomerism?

(A) CH_3 / CH_3 — C = C — H / CH_3

(B) CH_3 / Br — C = C — H / CH_2CH_3

(C) CH_3 / H — C = C — CH_3 / H

(D) Cl / H — C = C — Cl / H

(E) CH_3 / H — C = C — H / CH_2CH_3

7. Examine the reaction of cis 2-butene shown below:

H, CH_3, H, CH_3 (cis 2-butene) —Br_2→

a → H, Br, CH_3 / H, Br, CH_3 or Fischer projection: CH_3 / H — Br / Br — H / CH_3 (a)

b → H, CH, Br / Br, H, CH_3 or Fischer projection: CH_3 / Br — H / H — Br / CH_3 (b)

Which of the following statements concerning the reaction on the previous page is true?

I. (a) and (b) are enantiomers.

II. The products (a) and (b) are called the meso product.

III. The reaction is referred to as anti addition.

IV. The meso product could be obtained if we start out by using trans-2-butene.

(A) I only.
(B) I and III only.
(C) I, II, and III only.
(D) III and IV only.
(E) I, III, and IV only.

8. Which one of the following is a dextrorotatory compound?

(A)

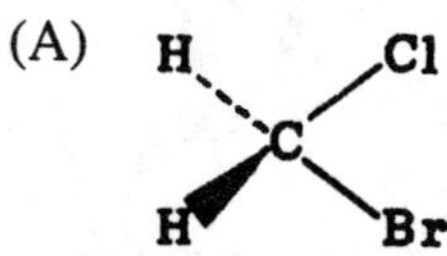

(B) CH_3 H

C

Br Cl

(C)

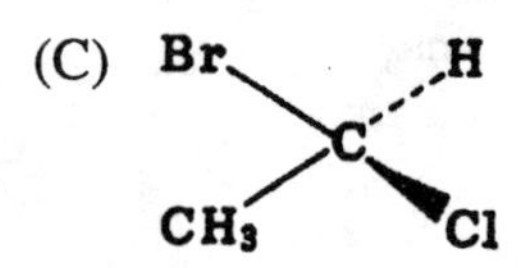

(D) None of the above.

(E) Cannot be decided by structure alone.

9. Consider the following Fischer projections:

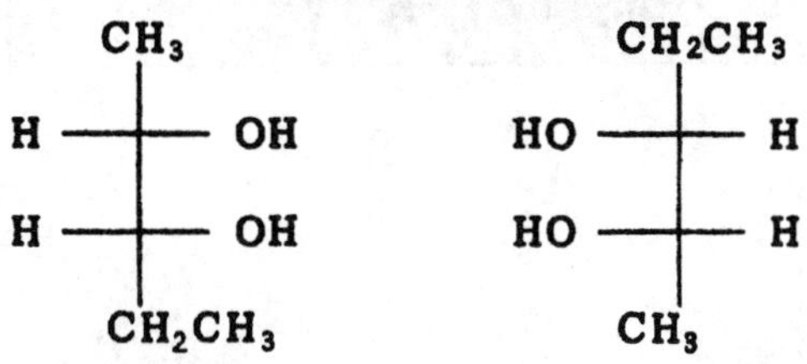

These compounds are

(A) identical.
(B) enantiomers.
(C) diastereomers.
(D) stereoselective.
(E) meso.

10. Which one of the following is NOT optically active?

(A)
CH_3
H — Cl
Br — Br
CH_3

(B)
CH_3
Cl — H
H — Cl
CH_3

(C)
CH_3
Br — H
H — Cl
CH_3

(D)
CH_3
H — Cl
H — Cl
CH_3

(E)
CH_3
Br — H
H — H
CH_3

ANSWER KEY

1. (C)
2. (D)
3. (D)
4. (D)
5. (B)
6. (A)
7. (D)
8. (E)
9. (A)
10. (D)

CHAPTER 7

Cyclic Hydrocarbons

Cyclic alkanes and cyclic alkenes are alicyclic (aliphatic cyclic) hydrocarbons.

7.1 Nomenclature

Cyclic aliphatic hydrocarbons are named by prefixing the term "cyclo-" to the name of the corresponding open-chain hydrocarbon, having the same number of carbon atoms as the ring.

Example

H_2C, CH_2, H_2C (three-membered ring)	H_2C—CH_2, H_2C—CH_2 (four-membered ring)	CH_2, H_2C, CH_2, H_2C—CH_2 (five-membered ring)
cyclopropane	cyclobutane	cyclopentane

Substituents on the ring are named, and their positions are indicated by numbers, the lowest combination of numbers being used.

Problem Solving Examples:

Give the IUPAC name for each of the following hydrocarbons.

(a) $(CH_3)_2CHCH_2CH_2CH(CH_3)_2$

(b)
```
                 CH3      CH3   CH2CH3
                 |        |     |
   CH3CH2CHCH2CCH2CHCH2CH3
                          |
                          CH3
```

(c)
```
              CH3CHCH2CH3
                  |
   CH3CH2CH2CHCH2CH3
```

(d)
```
          CH3CHCH3           CH2CH3
              |              |
   CH3CH2CHCH2CH2CCH2CH3
                             |
                             CH3
```

(e) Cyclopentane ring with two CH_3 groups on the same carbon

```
      CH3
   [ring]
      CH3
```

(f)
```
                 CH3
                 |
   [□]—CH2CHCH2CH2CH3
```

(g)
```
              CH3
              |
   (CH3CH2CCH2CH2CH2)3CH
              |
              CH3
```

(h) $(CH_3CH_2)_4C$

(i)
```
                  CH3
                  |
   (CH3CH2)2CHCHCH2CH3
```

(j)
```
                  CH3
                  |
   (CH3CH2)2CHCCH2CH3
                  |
                  CH3
```

(k) $(CH_3)_3C\overset{\triangledown}{C}HCH_3$

The naming of organic compounds by the IUPAC system follows certain procedures:

(1) Determine and identify by inspection the longest continuous carbon chain in the molecule or compound.

(2) Locate and name all the side chains attached to the longest continuous carbon chain.

(3) Indicate the names and positions and numbers of each type of side chain by means of prefixes to the name of the parent chain. The side chains are named in alphabetical order.

(4) Since cycloalkanes are basically saturated hydrocarbons, they are given the same name as their corresponding chain alkane, with the addition of the prefix cyclo- to signify the presence of a ring structure.

(a) We can rewrite the compound in its expanded form in order to make the structure more evident.

$$\begin{array}{c} \quad CH_3 \qquad\qquad CH_3 \\ \quad | \qquad\qquad\quad | \\ CH_3-CH-CH_2CH_2CH-CH_3 \end{array}$$

The longest continuous carbon chain consists of six carbon atoms, and one methyl group is attached to each of the two terminal carbons. The IUPAC name is thus 2,5-dimethylhexane.

(b)

$$\begin{array}{ccccccccc} & & & & CH_3 & & \overset{2}{C}H_2\overset{1}{C}H_3 & & \\ & & & & |^{5\ 4} & & ^{3}| & & \\ \overset{9}{C}H_3- & \overset{8}{C}H_2- & \overset{7}{C}H- & \overset{6}{C}H_2- & C- & CH_2- & CH- & CH_2- & CH_3 \\ & & | & & | & & & & \\ & & CH_3 & & CH_3 & & & & \end{array}$$

The longest chain contains nine carbon atoms. A methyl group is attached to the seventh carbon; two methyl groups are attached to the fifth carbon; and an ethyl group is bonded to the third carbon. The name of this compound is thus 3-ethyl-5,5,7-trimethylnonane.

(c)

```
                  3    2     1
            CH3-CH-CH2-CH3
                  4|
 7     6      5    |
CH3-CH2-CH2-CH-CH2-CH3
```

The longest chain has seven carbon atoms. The third carbon has a methyl group attached to it and the fourth carbon an ethyl group. The name given to this compound is 4-ethyl-3-methylheptane (alphabetical order has precedence over numerical order in side chains).

(d)

```
          CH3-CH-CH3        CH2CH3
 8     7     6/   5      4     3|  2     1
CH3-CH2-CH-CH2-CH2-C-CH2-CH3
                               |
                              CH3
```

Using the same principles illustrated previously, this compound is named 3-ethyl-6-isopropyl-3-methyloctane.

(e)

```
  1 CH3
 (ring)
    CH3
```

This structure is a five-carbon ring (cyclopentane) having two methyl groups attached to the same carbon (conventionally designated as C_1). The name of the structure is 1,1-dimethylcyclopentane.

(f)

```
          1     2CH3   3     4     5
 (ring)-CH2-CH-CH2-CH2-CH3
```

The longest chain has five carbon atoms. The ring is composed of four carbon atoms and is therefore cyclobutane. Thus, the name of the compound is 1-cyclobutyl-2-methylpentane.

(g) Written in expanded form:

```
                                        CH3
                                       3|
                            CH2CH2CH2CCH2CH3
                            |            |
            CH3             |           CH3
          11|              7|
CH3-CH2-CCH2CH2CH2-CH
            |               |
            CH3             |           CH3
                            |            |
                            CH2CH2CH2CCH2CH3
                                         |
                                        CH3
```

The longest chain has 13 carbon atoms (tridecane). The seventh carbon has one 4,4-dimethylhexyl group attached to it; in addition, two methyl groups are bonded to both the third and the eleventh carbon. The IUPAC name of the compound is written as 7-(4,4-dimethylhexyl)-3,3,11,11-tetramethyltridecane.

(j)

```
      CH3-CH2 CH3
            |  |
CH3-CH2-CHCCH2CH3
               |
              CH3
```

The longest chain has six carbons. The name of the compound is 3-ethyl-4,4-dimethylhexane.

(k)

```
  CH3  ▽
      \/
CH3-CCHCH3
    |
   CH3
```

The longest chain has four carbon atoms. The name of this compound is 2-cyclopropyl-3,3-dimethylbutane.

Write structural formulas for all of the possible cis-trans isomers of the following compounds:

(a) 1,2,3-trimethylcyclopropane

(b) 1,3-dichlorocyclopentane

(c) 1,1,3-trimethylcyclohexane

(d) (3-methylcyclobutyl)-3-methylcyclobutane

The first step undertaken in writing the structural formula for a compound is to draw the structure for the longest continuous carbon chain, which is indicated in the name. The next step is to attach the subgroups, denoted in the prefixes, to the parent chain at the positions indicated.

(a) 1,2,3-trimethylcyclopropane. Cyclopropane is planar and is a three-carbon ring (Δ) structure. Here, one methyl group is bonded to each carbon in the ring. Each methyl group can either be attached above the plane of the cyclopropane ring or below the plane. Only two different combinations of methyl group attachment are possible for this compound, resulting in two isomers:

CH_3 H_3C CH_3 CH_3 H_3C CH_3

The structural formulas can also be represented by

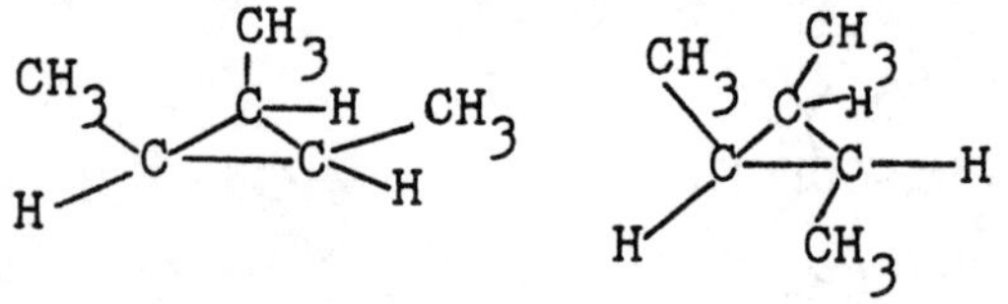

CH_3 CH_3 C—H CH_3 C C H H CH_3 CH_3 CH_3 C—H C C—H H CH_3

respectively, but we will use the first type of representation in our solutions due to its clarity.

(b) 1,3-dichlorocyclopentane. The last part of the name tells us that the compound is a five-carbon ring structure. Again, the substituents of the ring can be attached either above or below the ring, giving two isomers:

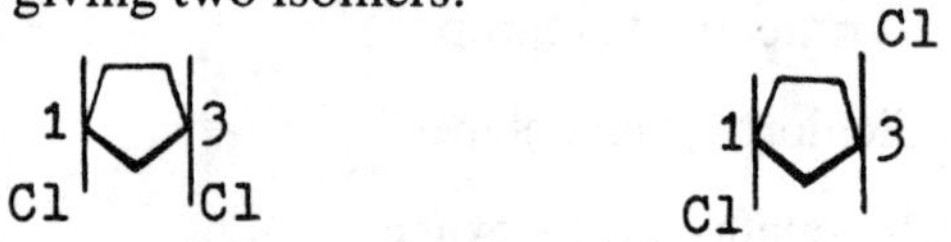

1 3 Cl Cl Cl 1 3 Cl

When both substituents are above or below the ring, they are cis with respect to the ring. If one is above and one is below the ring, they are trans.

(c) 1,1,3-trimethylcyclohexane. With some careful inspection one will notice that there is only one isomer present if the compound has a planar structure. However, because cyclohexane in reality is not planar, there are two conformational isomers of 1,1,3-trimethylcyclohexane:

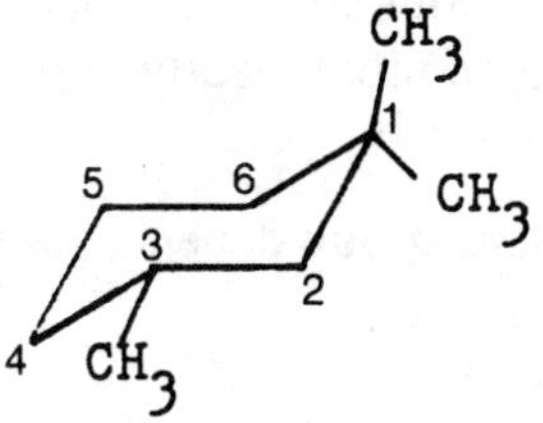

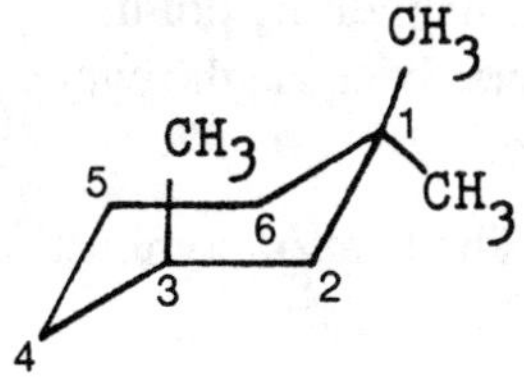

Equatorial $-CH_3$ Axial $-CH_3$

Since in room temperature these two forms undergo rapid interconversion, it is very difficult to distinguish them from each other.

(d) (3-methylcyclobutyl)-3-methylcyclobutane. A 3-methylcyclobutyl group is attached to the first carbon atom of 3-methylcyclobutane. Three possible arrangements of the two methyl groups exist:

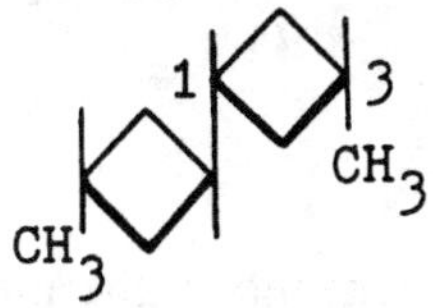

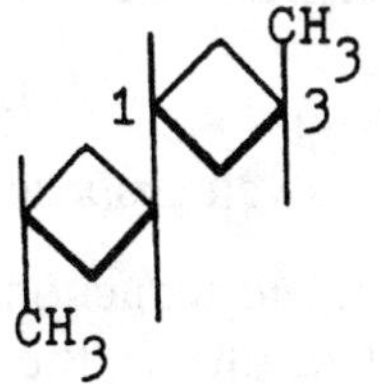

cis trans trans

Note that the two trans forms cannot be made to superimpose upon each other and are therefore not identical compounds.

7.2 Properties of Cyclic Hydrocarbons

Cyclopropane and cyclobutane are both colorless gases but cyclopentane is a colorless liquid.

The boiling points of the cycloalkanes are about 10°–20° higher than those of corresponding open-chain alkanes.

Their densities increase with an increase in the carbon chain length.

Cyclocylkanes and cycloalkenes are insoluble in water but soluble in alcohol and ether.

The melting points of cyclic hydrocarbons are higher than those of alkanes because the cyclic hydrocarbons fit more readily into a crystal lattice.

The heat of combustion per methylene group depends on the ring size.

7.3 Preparation of Cyclic Hydrocarbons

Reaction of active metals (Na, Mg, Zn, etc.) with certain dihalogenated hydrocarbons.

a) $X - (CH_2)_n - X + 2M(\text{or } M) \rightarrow (CH_2)_n + 2MBr\ (\text{or } MBr_2)$

cycloalkane

Example

$$BrCH_2CH_2CH_2Br + Zn \rightarrow H_2C(CH_2)CH_2\text{ (ring)} + ZnBr_2$$

cyclopropane

b) Partial or complete reduction of benzene in the presence of heated nickel results in the formation of cyclohexadiene, cyclohexene, or cyclohexane.

Example

$$C_6H_6 + 3H_2 \xrightarrow[\text{Heat } < 300°C]{\text{finely divided Ni}} C_6H_8 \rightarrow C_6H_{10} \rightarrow C_6H_{12}$$

$$C_6H_5CH_3 + 3H_2 \xrightarrow{Ni/Pt/Pd} C_6H_{11}CH_3$$

Hydrogenation of Arenes

Addition of carbenes to alkenes

$$>C=C< \; + :CX_2 \rightarrow \text{(cyclopropane ring: } CX_2 \text{ bridging } -C-C-) \qquad X = H, Cl, Br$$

Problem Solving Example:

What would be the expected products from the reaction between dichlorocarbene and the following alkenes?

(a) 2-methylpropene

(b) cyclohexene

(c) 3-methylpentene

Dichlorocarbene is a reactive intermediate that will add to double bonds to form cyclic hydrocarbons.

(a) $CH_3-C(CH_3)=CH_2 + :CCl_2 \longrightarrow CH_3-C(CH_3)-CH_2$ (ring closed through CCl_2: C bearing Cl, Cl)

(b) cyclohexene $+ :CCl_2 \longrightarrow$ 7,7-dichlorobicyclo (Cl, Cl)

(c) methylcyclopentene (CH_3) $+ :CCl_2 \longrightarrow$ (CH_3; Cl Cl)

7.4 Reactions of Cycloalkanes and Cycloalkenes

Reaction of bromine with cyclopentane to give bromocyclopentane.

Example

$$\begin{matrix} CH_2-CH_2 \\ | \\ CH_2-CH_2 \end{matrix} \!\!> CH_2 + Br_2 \xrightarrow{300°C} \begin{matrix} CH_2CH_2 \\ | \\ CH_2CH_2 \end{matrix} \!\!> CHBr + HBr$$

bromocyclopentane

Cycloalkanes undergo free-radical substitution.

Example

$$\text{cyclo-}(H_2C{-}CH_2{-}H_2C) + Cl_2 \xrightarrow[\text{heat}]{\text{light}} \text{cyclo-}(H_2C{-}CHCl{-}H_2C) + HCl$$

cyclopropane chlorocyclopropane

Cycloalkenes undergo both electrophilic and free-radical addition reactions; they also undergo cleavage and allylic substitutions.

$$\text{Cyclohexene} + Cl_2 \rightarrow \text{1,2-Dichlorocyclohexane}$$

$$\text{1-Methylcyclopentene} + HBr \rightarrow \text{1-Bromo-1-methylcyclopentane}$$

$$\text{3,5-Dimethylcyclopentene} \xrightarrow{O_3,\ H_2O/Zn} O{=}CH{-}CH(CH_3){-}CH_2{-}CH(CH_3){-}CH{=}O$$

A dialdehyde

Chain addition reactions of cyclopropane and cyclobutane: These addition reactions destroy the cyclopropane and cyclobutane ring systems and yield open-chain products.

$$\text{cyclo-}(H_2C{-}CH_2{-}H_2C) \xrightarrow{Ni,\ H_2,\ 80°C} CH_2(H){-}CH_2{-}CH_2(H) \text{ (Propane)}$$

$$\text{cyclo-}(H_2C{-}CH_2{-}H_2C) \xrightarrow{Cl_2,\ FeCl_3} CH_2(Cl){-}CH_2{-}CH_2(Cl) \text{ (1,3-Dichloropropane)}$$

$$\text{cyclo-}(H_2C{-}CH_2{-}H_2C) \xrightarrow{\text{Conc. } H_2SO_4} CH_2(H){-}CH_2{-}CH_2(OH) \text{ (1-Propanol)}$$

Problem Solving Example:

Q Starting with any organic compound of less than four carbon atoms and any inorganic or physical reagent that may be required, synthesize the following:

(a) 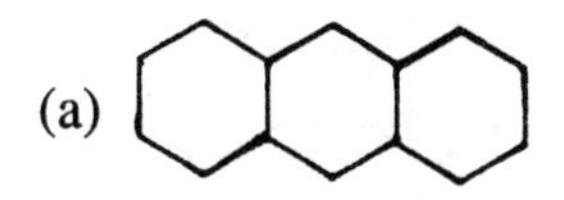

(b) 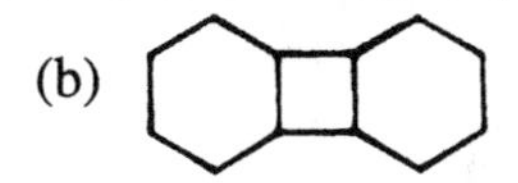

(c) 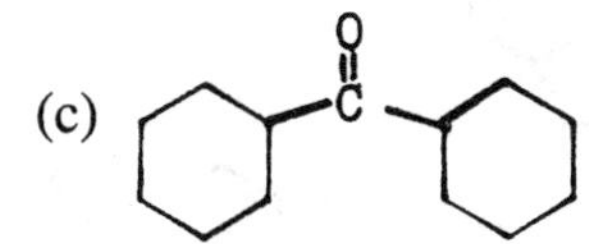

A Diels-Alder reactions, as well as other electrocyclic reactions, are very valuable synthetic tools. They enable the chemist to perform ring closings, build up molecules, and are used extensively in the syntheses of drugs.

(a) 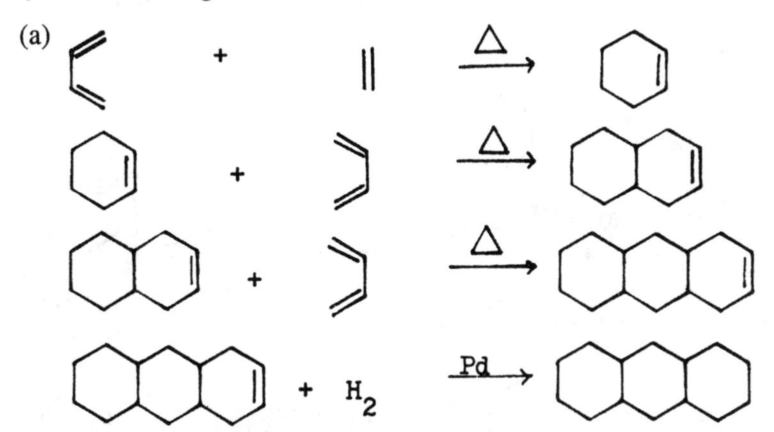

(b) This reaction is a slight variation on the Diels-Alder. The first step consists of the formation of cyclohexene.

Now, if two cyclohexenes are exposed to ultraviolet light, the desired product will be obtained.

$$\text{cyclohexene} + \text{cyclohexene} \xrightarrow{h\nu} \text{(fused tricyclic cyclobutane dimer)}$$

(c) This reaction involves little more than cyclohexane formation followed by two Grignard reactions.

$$\text{butadiene} + \text{ethylene} \xrightarrow{\Delta} \text{cyclohexene}$$

$$\text{cyclohexene} + HCl \xrightarrow{Pt} C_6H_{11}Cl \xrightarrow{Mg} C_6H_{11}MgCl$$

$$C_6H_{11}MgCl + HCHO \longrightarrow C_6H_{11}CH_2OMgCl \xrightarrow{H_2O} C_6H_{11}CH_2OH$$

$$\xrightarrow{Cr_2O_7} C_6H_{11}CHO$$

$$C_6H_{11}CHO + ClMgC_6H_{11} \longrightarrow C_6H_{11}CH(OMgCl)C_6H_{11}$$

$$\xrightarrow{H_2O} C_6H_{11}CH(OH)C_6H_{11} \xrightarrow{H_2Cr_2O_7} C_6H_{11}C(=O)C_6H_{11}$$

7.5 Baeyer Strain

When carbon is bonded to four other atoms, the angle between any pair of bonds is a tetrahedral angle of 109.5°. But the ring of cyclopropane is a triangle with three angles of 60°. These deviations of bond angles from the "normal" value cause the molecule to be strained and hence unstable. The more unstable the molecule is the more prone it is to undergo ring opening reactions.

Problem Solving Example:

Q (a) Estimate the amount of eclipsing strain in planar cyclopentane. (b) Is the amount of eclipsing strain decreased in going from the planar to the puckered form?

A Eclipsing strain has a value of approximately 1 Kcal/mole for two adjacent, eclipsed bonds. In planar cyclopentane, 10 pairs of eclipsed hydrogens are found, thus 10 Kcal of eclipsing strain is present.

As planar cyclopentane folds, the number of eclipsed hydrogens decreases, and thus eclipsing strain decreases.

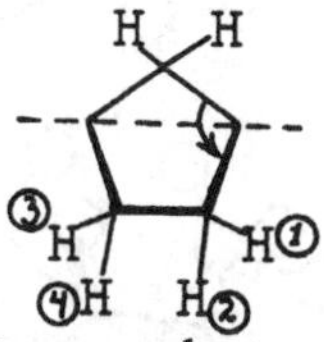

H H

③H H①

④H H②

In puckered cyclopentane, only two pairs of eclipsed hydrogens are found: 1 and 3, 2 and 4. Hence, only 2 Kcal of eclipsing strain is present.

Therefore, as planar cyclopentane becomes puckered, a loss of 8 Kcal of eclipsing strain enables the puckered form to become more stable.

7.6 Conformation of Cycloalkanes

Factors Affecting Stability of Conformations

An angle strain (Baeyer strain) accompanies deviations from the "normal" bond angles (tetrahedral, 109.5° angle).

Carbon atoms in a tetrahedral arrangement tend to assume a staggered conformation.

A torsional strain (Pitzer strain) accompanies deviations from staggered arrangement.

The distance between nonbonding orbitals is equal to the sum of their van der Waals radii. This causes them to attract each other.

When the attracted atoms are brought closer together, they repel each other, and a van der Waals strain (steric strain) results.

Nonbonded atoms (or groups of atoms) tend to take positions that result in the most favorable dipole-dipole interactions.

A molecule accepts a certain amount of angle strain to relieve van der Waals strain or dipole-dipole interactions.

Conformations of Cycloalkanes

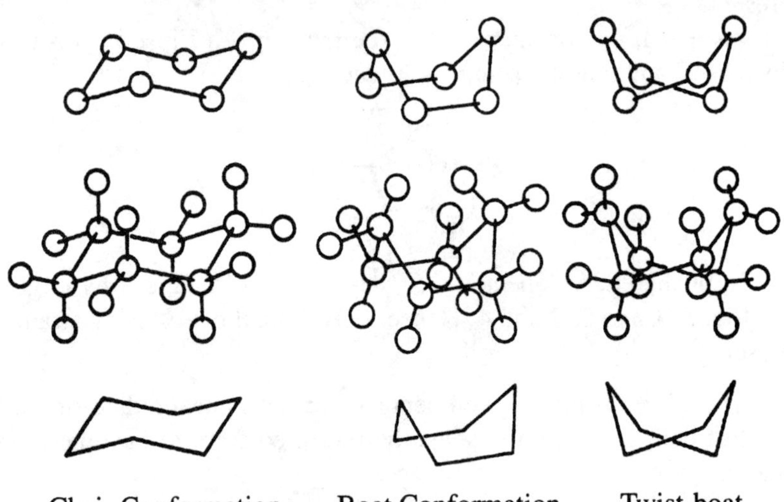

Chair Conformation | Boat Conformation (an energy maximum) | Twist-boat Conformation

Conformations of cyclohexane that are free of angle strain.

Along each of the carbon-carbon bonds in the chair form, there are perfectly staggered bonds.

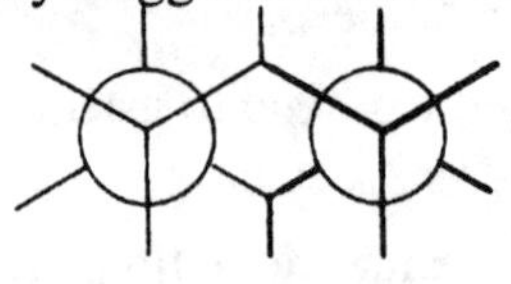

Chair cyclohexane

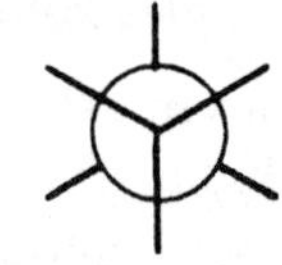

Staggered ethane

The chair form is the most stable conformation of cyclohexane and of nearly every derivative of cyclohexane. The chair form is a conformational isomer since it lies at an energy minimum.

Along each of the two carbon-carbon bonds in the boat conformation, there are sets of exactly eclipsed bonds.

Flagpole bonds

Boat cyclohexane

Eclipsed ethane

The chair conformation is much more stable than the boat conformation. The boat conformation is not a conformer, but a transition state between two conformers since it lies at an energy maximum.

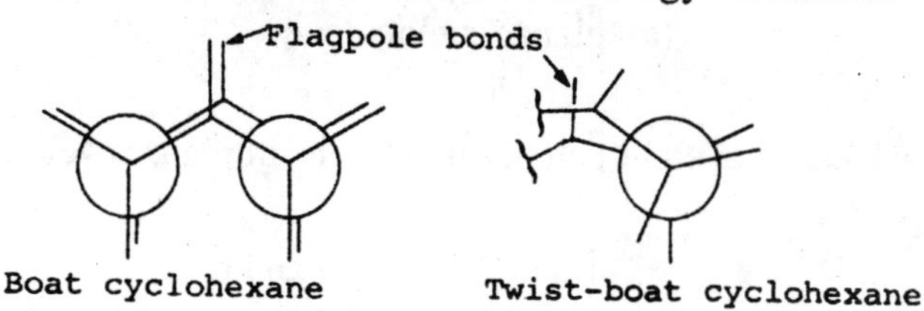

The twist-boat conformation is a conformer since it lies at an energy minimum at 5.5 Kcal above the chair conformation.

Equatorial and Axial Bonds in Cyclohexane

The equatorial bonds that hold the hydrogen atoms in the plane of the ring lie about the "equator" of the ring. The axial bonds that hold the hydrogen atoms above and below the plane are pointed along an axis perpendicular to the plane.

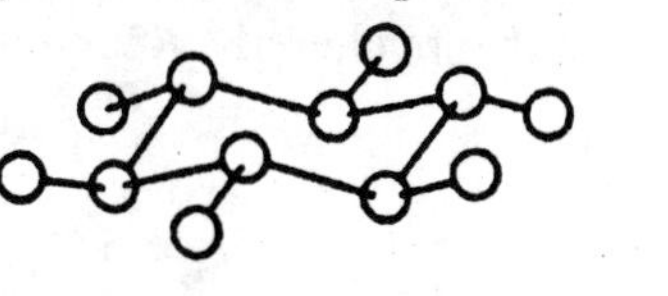

Axial Bonds

1,3 diaxial interaction results from the severe crowding among atoms that are held by the three axial bonds on the same side of the molecule. A given atom or group, except for hydrogen, has less room in an axial position than in an equatorial position.

Problem Solving Examples:

(a) How many pairs of eclipsed hydrogens are present in planar cyclobutane? (b) In puckered cyclobutane?

First, let us look at the structure of planar cyclobutane.

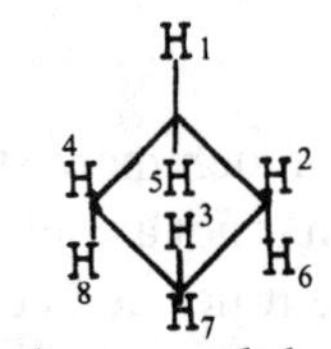

(a) planar cyclobutane

As we can see, eight pairs of hydrogens in planar cyclobutane are eclipsed:

1 and 2	5 and 6
2 and 3	6 and 7
3 and 4	7 and 8
4 and 1	8 and 5

In puckered cyclobutane, we can tell from examining its structure that no eclipsed hydrogens are present.

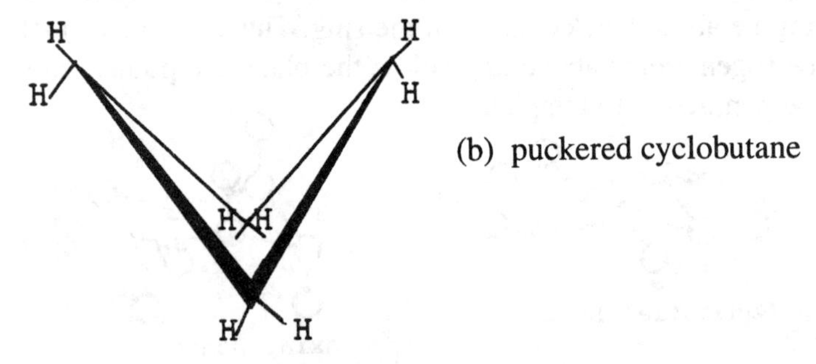

(b) puckered cyclobutane

The student should use a model if this is not clear.

Would you expect cis- or trans-1,2-dimethylcyclopropane to be the more stable? Explain.

The trans form is expected to be more stable than the cis form. As we can see, the former has its methyl groups farther apart from each other than the latter. The nonbonded interaction between the hydrogen and the methyl group in the trans conformation is smaller than that between the two methyl groups in the cis conformation. Thus, the cis form has a greater strain and is less stable.

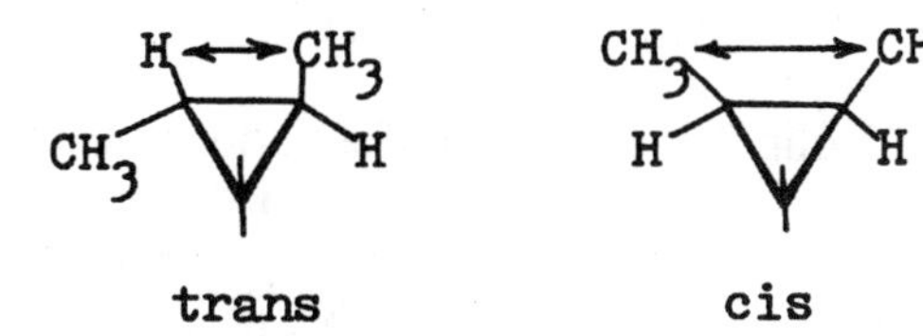

7.7 Stereoisomerism of Cyclic Compounds

Stereoisomerism of cyclic compounds: cis- and trans-isomers.

cis-1,2-cyclopentanediol trans-1,2-cyclopentanediol

The above isomers are configurational isomers; they are isolable because they are interconverted only by the breaking of bonds.

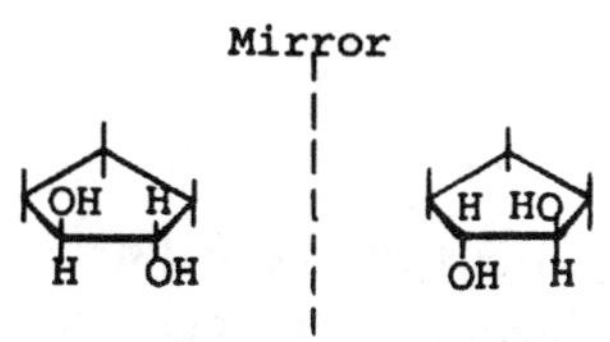

Not superimposable
Enantiomers resolvable
trans-1,2-cyclopentanediol

The trans-glycol is chiral, and its isomers are resolvable into optically active compounds (enantiomers).

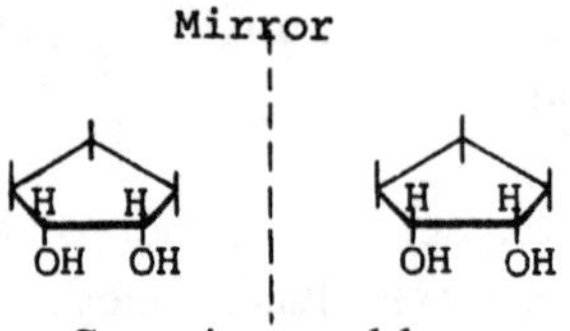

Mirror

Superimposable
A meso compound

cis-1,2-cyclopentanediol

The cis-glycol is a meso compound. By definition its mirror images are identical and unresolvable. The compound is not chiral and is optically inactive.

Stereoisomerism of Cyclic Compound Conformational Analysis

Diequatorial Diaxial

Chair conformations of trans-1,2-dimethylcyclohexane.

The diequatorial conformation is the more stable one because there is less crowding between – CH_3 groups and axial hydrogens of the ring (less 1,3 diaxial interaction).

Equatorial-axial Axial-equatorial

Chair conformations of cis-1,2-dimethylcyclohexane.

The two cis- conformations are of equal stability and are less stable than the trans- isomer. Since in either cis- conformation, one –CH_3 group has to be in the axial position (1,3-diaxial interaction).

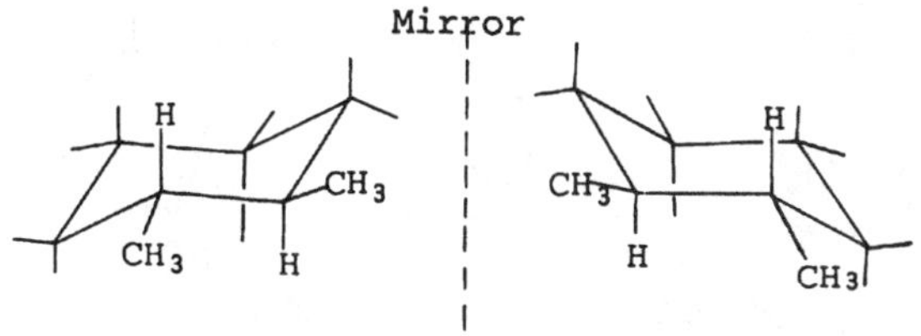

Not superimposable; not interconvertible
trans-1,2-dimethylcyclohexane

A resolvable racemic modification

The trans-1,2-dimethylcyclohexane mirror images are not superimposable and therefore are enantiomers; they are not interconvertible and could be resolved into the enantiomers, each of which is optically active.

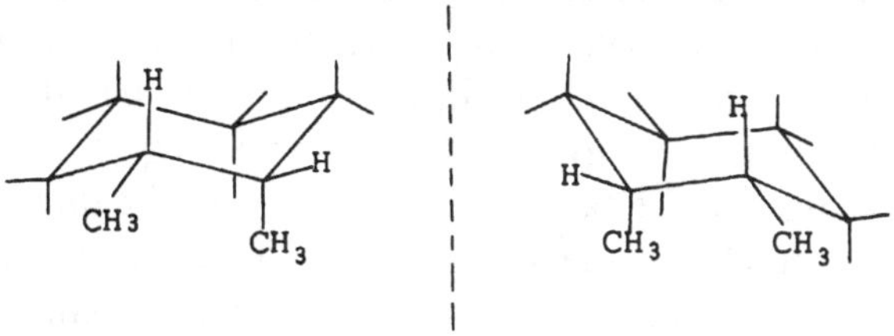

Not superimposable; but interconvertible
cis-1,2-dimethylcyclohexane

A non-resolvable racemic modification

The cis-1,2-dimethylcyclohexane mirror images are not superimposable, and therefore, are enantiomers. In this case the enantiomers are interconvertible and cannot be resolved. These are conformational enantiomers.

Problem Solving Examples:

Q You have two bottles labeled "1,2-cyclopentanediol," one containing a compound of m.p. 30°, the other a compound of m.p. 55°; both compounds are optically inactive. How could you decide, beyond any doubt, which bottle should be labeled "cis" and which "trans"?

A

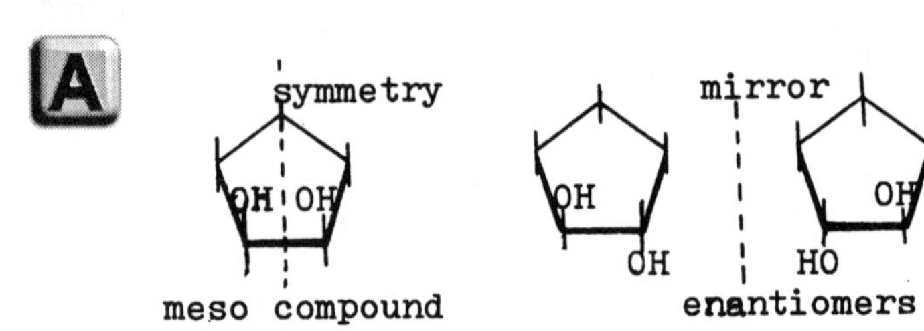

The mirror image of cis-1,2-cyclopentanediol is identical to and superimposable upon the cis compound itself. By examining the structure a bit more closely, we find that the compound is actually a meso compound. By definition, its mirror images are identical and so unresolvable, and the compound is optically inactive.

Trans-1,2-cyclopentanediol has nonsuperimposable mirror images, and so can exist as two enantiomers. When equal parts of enantiomers are mixed, trans-1,2-cyclopentanediol is a racemic modification and appears to be optically inactive. This is because the equal parts of enantiomers rotate light to the same degree but in opposite directions, and the net effect is no observable optical activity. However, the trans compound can be resolved into its enantiomers by special methods, such as reacting them with reagents that are themselves optically active.

Thus, if the solution is resolvable into two optically active compounds, it must contain the trans isomers.

Q Which of the following compounds are resolvable, and which are nonresolvable? Which are truly meso compounds? Use models as well as drawings.

(a) cis-1,2-cyclohexanediol
(b) trans-1,2-cyclohexanediol
(c) cis-1,3-cyclohexanediol
(d) trans-1,3-cyclohexanediol
(e) cis-1,4-cyclohexanediol
(f) trans-1,4-cyclohexanediol

A Every compound has a mirror image. But not all mirror images are superimposable on each other. Mirror images that are nonsuperimposable are called enantiomers. Enantiomers can be resolved and are characterized by optical activity. A compound with a chiral center (atom bonded to four different groups) has enantiomers, whereas one with no chiral center has no enantiomers. However, there are compounds with chiral carbons whose mirror images are superimposable. These compounds are called meso compounds. Meso compounds cannot be resolved.

(a) cis-1,2-cyclohexanediol

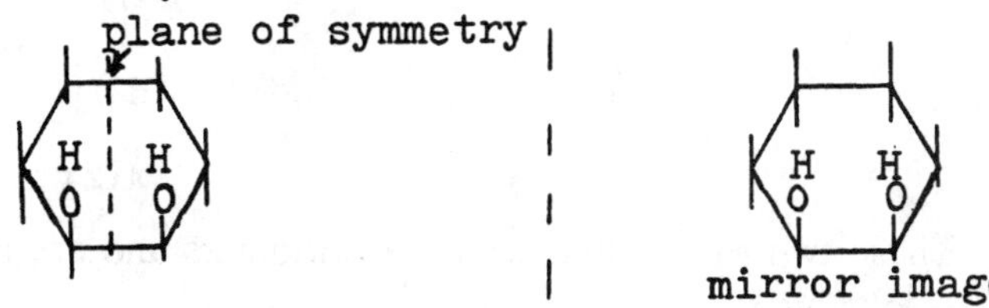

These mirror images are actually identical and superimposable on each other. On closer inspection one-half of the compound is superimposable on its other half. This is equivalent to saying that there is a plane of symmetry in the compound. Cis-1,2-cyclohexanediol is a meso compound and is not resolvable, although it contains two chiral carbons.

(b) trans-1,2-cyclohexanediol

OH
OH
OH
HO
mirror image

These mirror images are not superimposable. Note also that there is no internal plane of symmetry in (b). Trans-1,2-cyclohexanediol, therefore, exists as enantiomers which are resolvable.

(c) cis-1,3-cyclohexanediol

OH
OH
symmetry
HO
OH
mirror image

These two forms are superimposable. They are actually identical. In addition, an internal symmetry exists, making cis-1,3-cyclohexanediol a meso compound. Meso compounds are not resolvable.

(d) trans-1,3-cyclohexanediol

mirror image

These two mirror images are enantiomers and are therefore resolvable.

(e) cis-1,4-cyclohexanediol

symmetry

mirror image

These two compounds are perfectly superimposable and thus nonresolvable. They are actually identical compounds.

(f) trans-1,4-cyclohexanediol

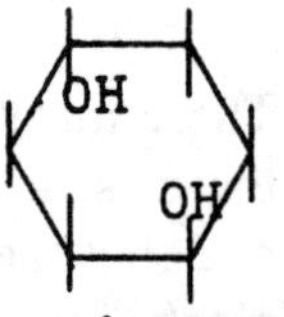

mirror image

Like compound (e), these are identical compounds and not enantiomers. They cannot be resolved.

Note that (e) and (f) are not meso compounds since they contain no chiral carbons.

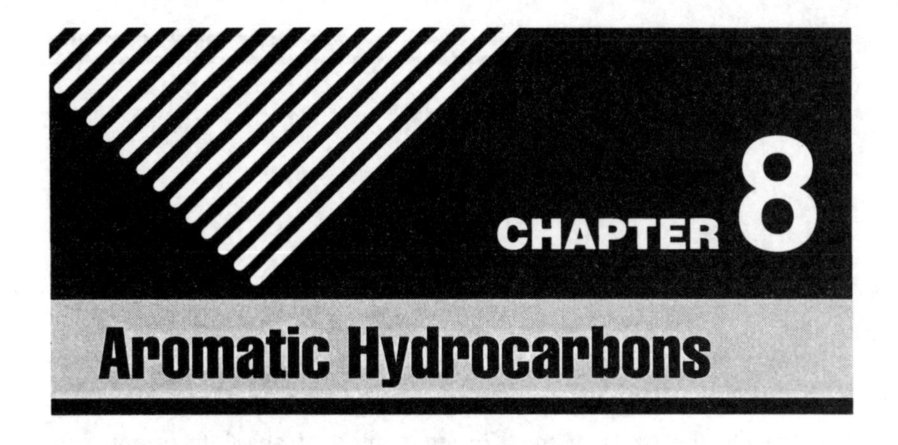

Aromatic Hydrocarbons

Most aromatic hydrocarbons (arenes) are derivatives of benzene. Examples of benzene derivatives are naphthalene, anthracene, and phenanthrene.

8.1 Structure

Benzene has a symmetrical structure and the analysis, synthesis, and molecular weight determination indicate a molecular formula of C_6H_6.

Naphthalene structure is indicated by the oxidation of 1-nitronaphthalene, which shows that the substituted ring is a true benzene ring. Reduction and oxidation of the same nucleus indicates that the unsubstituted ring is a true benzene ring.

COOH, COOH ← oxidation ← NH_2 (1, 2) ← reduction ← NO_2 (1, 2) → oxidation → HOOC, HOOC, NO_2

Problem Solving Examples:

Q For a time the prism formula VI, proposed in 1869 by Albert Ladenburg of Germany, was considered as a possible structure for benzene, on the grounds that it would yield one monosubstitution product and three isomeric disubstitution products.

```
CH———CH
|  \   / |
|   CH   |
|   |    |
CH——+——CH
   \ | /
    CH
```

VI

(a) Draw Ladenburg structures of three possible isomeric dibromobenzenes.

(b) On the basis of the Körner method of absolute orientation, label each Ladenburg structure in (a) as ortho, meta, or para.

(c) Can the Ladenburg formula actually pass the test of isomer number? (Derivatives of Ladenburg "benzene," called prismanes, have actually been made.)

(a) The three isomeric Ladenburg structures for dibromobenzenes are:

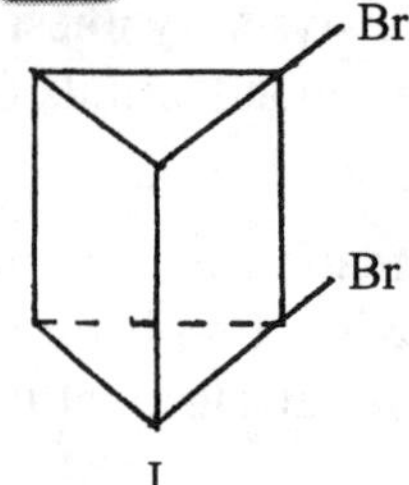

I

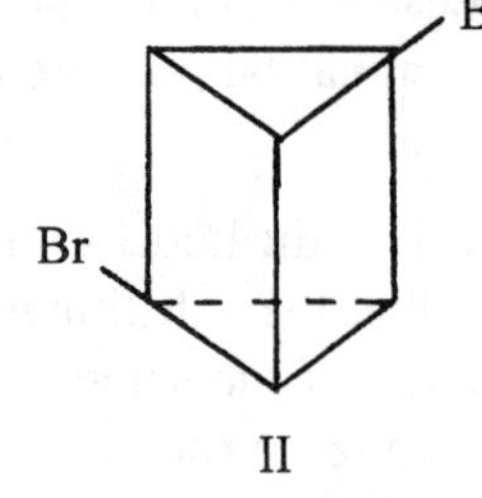

II

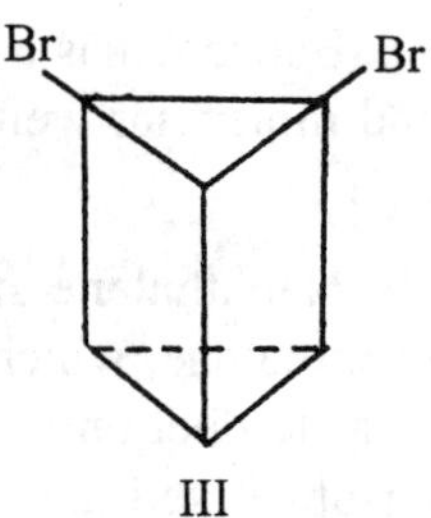

III

These three structures are different isomers because isomer I has the two bromines between the edges of two squares; isomer II has one bromine on the corner of one triangle, and the other bromine on the corner of the other triangle; isomer III has the two bromines on the two corners of the same triangle. These are the only three possible isomeric Ladenburg dibromobenzenes.

(b) According to the Körner method of absolute orientation, the dibromobenzene isomer that reacts to form one isomer of dibromonitrobenzene is the para-dibromobenzene; the one that forms two isomers is the ortho-dibromobenzene; and the one that forms three isomers is the meta-dibromobenzene. By drawing the possible isomeric dibromonitrobenzenes for each isomer of dibromobenzene, we can label each Ladenburg structure in (a) as ortho, meta, or para:

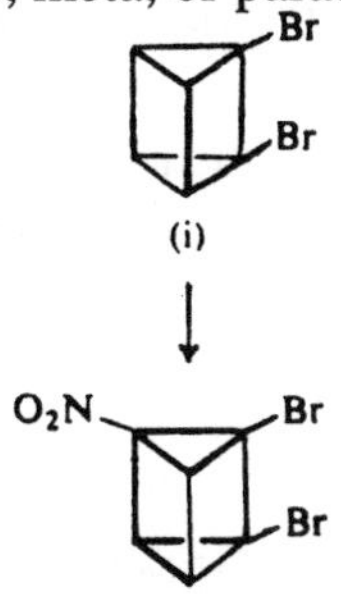

(i)

One

Since there is only one possible isomeric product, this dibromobenzene is the para isomer.

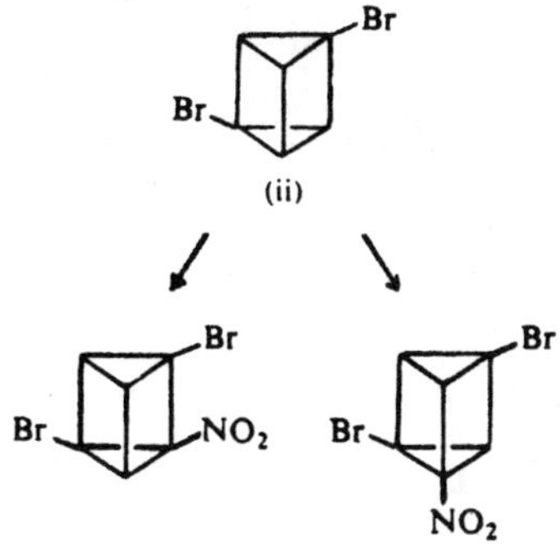

(ii)

Two

There are two isomeric products; therefore, this is ortho-dibromobenzene.

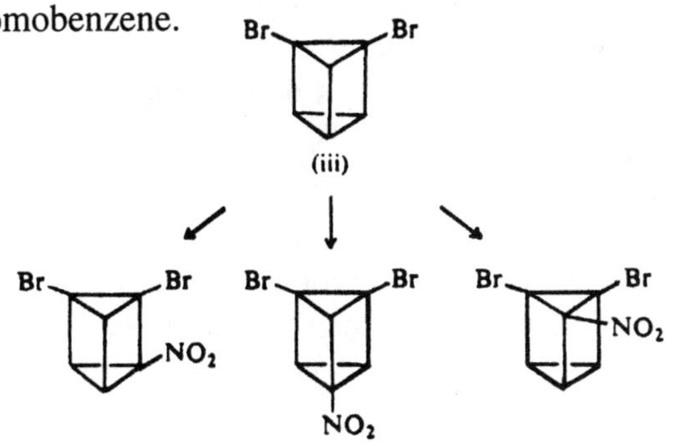

(iii)

Three

There are three isomeric products; therefore, this is meta-dibromobenzene.

(c) No. All three disubstituted benzenes are not chiral. For the Ladenburg formula, the ortho isomer is chiral; thus, enantiomeric structures are possible. (Recall that molecules that are not superimposable on their mirror images are chiral and that enantiomers are mirror-image isomers.):

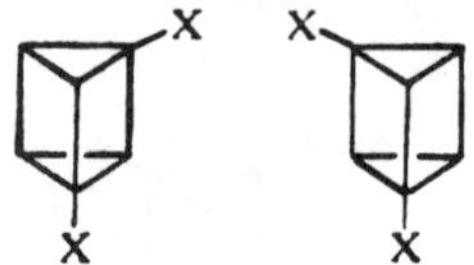

Enantiomers

Q In Kekulé's day, one puzzling aspect of his dynamic theory for benzene was provided by 1,2-dimethylbenzene. According to his theory, there should be two distinct such compounds, one with a double bond between the two methyl-substituted carbons and one with a single bond in this position

CH_3 CH_3

CH_3 CH_3

Only a single 1,2-dimethylbenzene is known, however.

(a) Does Ladenburg's formula solve this problem?

(b) Explain with modern resonance theory.

(a) Ladenburg's structure for benzene is a tetracyclic one (notice the absence of double bonds):

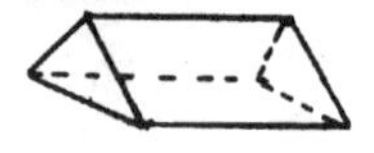

Fig. 1

This structure accounts for the fact that benzene has only one monosubstitution product and three isomeric disubstitution products. However, Ladenburg's formula is not able to solve the problem of there being only one 1,2-dimethylbenzene. According to this formula, there would be two possible isomers for this compound:

Fig. 2

The two methyl groups in isomer I are shared by the edges of two squares, while those of isomer II are shared by the edges of a square and a triangle. Since the methyl groups in these two structures have different spatial orientations, they are isomers of each other, thereby contradicting the experimental observation of one isomer.

(b) Modern resonance theory says that the π electrons are delocalized in a π cloud. A pictorial representation of the π cloud in the benzene ring is:

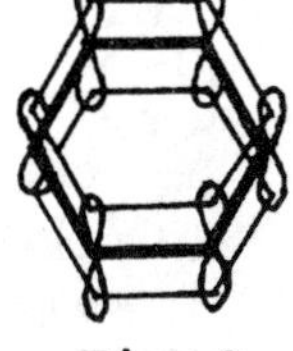

Fig. 3

The thin lines depict delocalization of the six π electrons. One structural representation of this system is the Kekulé's structure (shown in Figure 4).

CH_3 CH_3 (a) ⟷ CH_3 CH_3 (b)

Fig. 4

The double-headed arrow and the brackets are used to indicate that the two structures are resonance structures. Remembering the concept of resonance (that is, the delocalization of electrons), we can see why these two structures are equivalent. Using the arrow convention of moving electrons, we have:

CH_3 CH_3 ↔ CH_3 CH_3 ≡ CH_3 CH_3

(a) (b) (c)

Fig. 5

These two resonance structures (Figures 5a and 5b) can also be represented by Figure 5c, with the circle representing the cloud of six delocalized π electrons.

8.2 Nomenclature (IUPAC System)

Aromatic compounds are named as derivatives of the corresponding hydrocarbon nucleus.

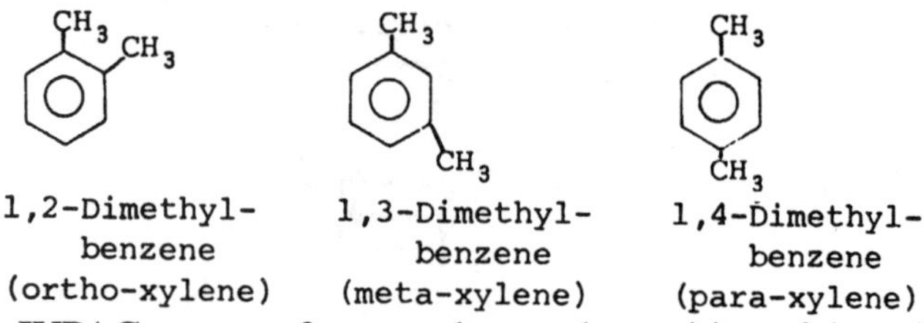

In the IUPAC system of nomenclature, the position of the substituent group is always indicated by numbers arranged in a certain order:

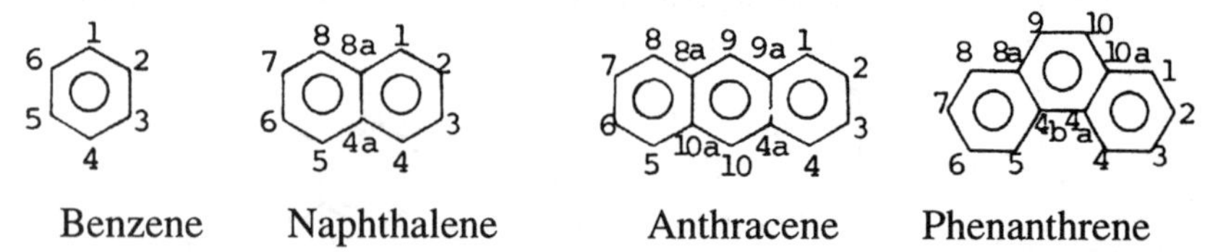

Problem Solving Example:

Draw structures of:

(a) p-dinitrobenzene
(b) m-bromonitrobenzene
(c) o-chlorobenzoic acid
(d) m-nitrotoluene
(e) p-bromoaniline
(f) m-iodophenol
(g) mesitylene(1,3,5-trimethylbenzene)
(h) 3,5-dinitrobenzenesulfonic acid
(i) 4-chloro-2,3-dinitrotoluene
(j) 2-amino-5-bromo-3-nitrobenzoic acid
(k) p-hydroxybenzoic acid
(l) 2,4,6-trinitrophenol (picric acid)

A All of these structures are derivatives of benzene. Hence, a consideration of the nomenclature of benzene derivations will aid in the drawing of these compounds.

For many derivatives, the name of the substituent is prefixed to the word "-benzene," as in iodobenzene.

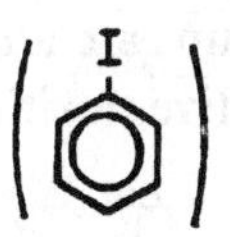

Others possess special names that show no resemblance to the name of the attached substituent group. For example, methylbenzene is termed toluene. The most important compounds in this class include:

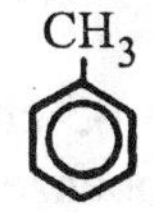

Toluene

Aniline

Phenol

Benzoic acid

Benzenesulfonic acid

When several groups are attached to the benzene ring, the positions as well as the names of the substituents must be indicated. The words ortho (o), meta (m), and para (p) are used to designate the three possible isomers of a disubstituted benzene. For example,

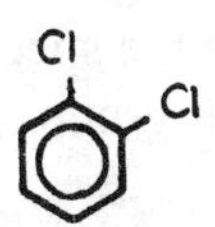

o-dichlorobenzene
ortho

m-dichlorobenzene
meta

p-dichlorobenzene
para

If the two groups are different, and neither group imparts a special name to the compound, then the two groups are named successively and end the word with -benzene. For example:

m-chloronitrobenzene p-bromoiodobenzene

If one of the two groups does impart a special name, then the compound is named as a derivative of that special compound. For example:

m-nitrobenzoic acid o-nitrotoluene

When more than two groups are attached to the benzene, numbers are used to indicate their relative positions. For example:

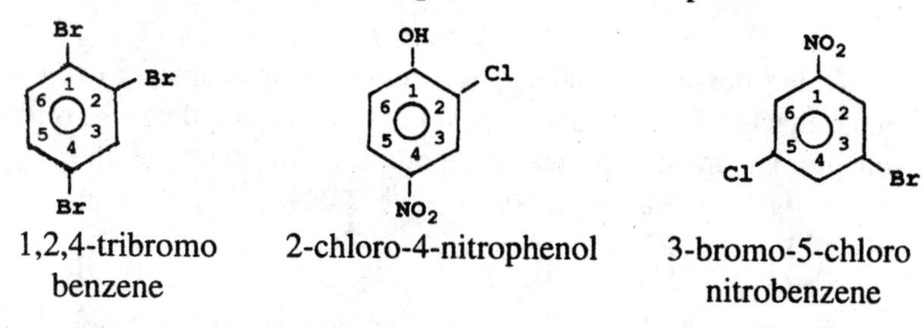

1,2,4-tribromo benzene 2-chloro-4-nitrophenol 3-bromo-5-chloro nitrobenzene

When the groups are all the same, each is given a number; the sequence being the one that gives the lowest combination of numbers. When the groups are different, the last named group is understood to be in position 1 and the other numbers conform to that, as in 3-bromo-5-chloro-nitrobenzene. When one of the groups that gives a special name exists, then the compound is named as having the special group in position 1; thus, in 2,6-dinitrotoluene the methyl group is considered to be at the 1-position.

From their names, the chemical structure of the given compounds can now be determined.

(a) p-dinitrobenzene

NO_2 — benzene — NO_2

(b) m-bromonitrobenzene

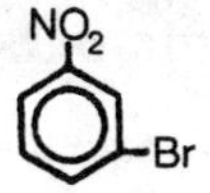

NO_2, Br

(c) o-chlorobenzoic acid

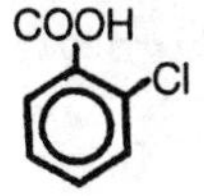

COOH, Cl

(d) m-nitrotoluene

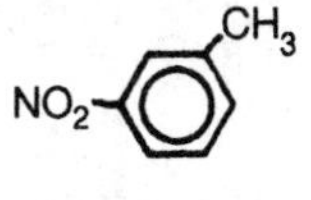

CH_3, NO_2

(e) p-bromoaniline

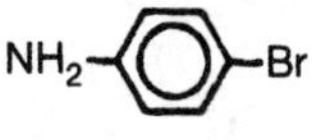

NH_2 — benzene — Br

(f) m-iodophenol

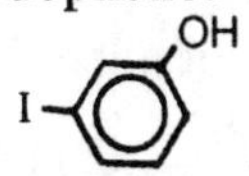

OH, I

(g) mesitylene

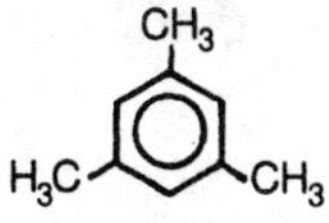

CH_3, H_3C, CH_3

(h) 3,5-dinitrobenzenesulfonic acid

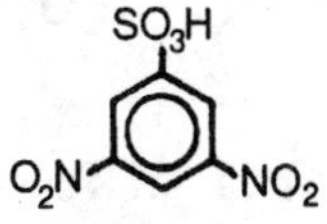

SO_3H, O_2N, NO_2

(i) 4-chloro-2,3-dinitro-toluene

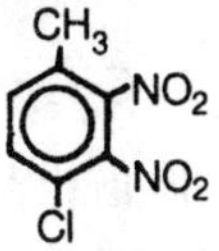

CH_3, NO_2, NO_2, Cl

(j) 2-amino-5-bromo-3-nitro benzoic acid

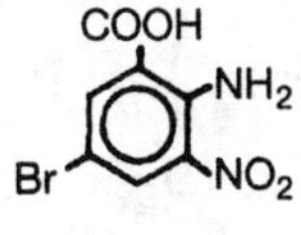

COOH, NH_2, Br, NO_2

(k) p-hydroxybenzoic acid

COOH, OH

(l) 2,4,6-trinitrophenol

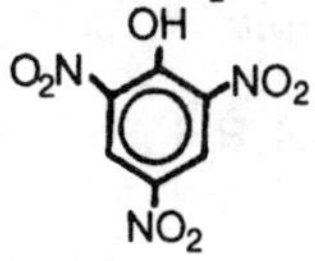

OH, O_2N, NO_2, NO_2

8.3 Preparation of Benzene and Its Derivatives

Preparation of Benzene

Passage of acetylene and alkanes (C_2H_2 to C_6H_{14}) through hot tubes.

Example

$$3H-C \equiv C-H \xrightarrow[580°C]{\text{through tube}} C_6H_6 \text{ (high yield)}$$

$$CH_3-CH_2-CH_2-CH_2-CH_2-CH_3 \xrightarrow[Cr_2O_3]{\text{hot tower}} C_6H_6$$

Heating of phenol with zinc dust.

$$C_6H_6OH + Zn\text{, dust} \rightarrow C_6H_6 + ZnO$$

Hydrolysis of benzene sulfonic acid with superheated steam.

$$C_6H_5SO_2OH + H_2O\text{, superheated steam} \xrightarrow[\text{catalyst}]{} C_6H_6 + H_2SO_4$$

Preparation of Toluene

The Wurtz Reaction—the action of sodium on a mixture of halobenzene and methylhalide.

Example $C_6H_5Br + 2Na + Br - CH_3$, ether $\rightarrow C_6H_5CH_3 + 2NaBr$

The Friedel-Crafts Reaction—action of an alkylhalide on benzene in the presence of anhydrous aluminum chloride.

Example

$$C_6H_5H + Br - CH_3 \xrightarrow[AlCl_3]{\text{anhydrous}} C_6H_5 - CH_3 + HBr$$

Preparation of Xylene

Reaction of a methyl halide with toluene in the presence of anhydrous aluminum chloride (Friedel-Crafts Reaction).

$$C_6H_5CH_3 + Br-CH_3 \xrightarrow{AlCl_3} \text{o-Xylene or p-Xylene} + HBr$$

o-Xylene

p-Xylene

Preparation of Naphthalene

Passage of benzene and acetylene through a hot tube.

$$C_6H_6 + 2H{-}C{\equiv}C{-}H \xrightarrow{\text{hot tube}} C_{10}H_8 + H_2$$

Reaction of zinc or sodium with a mixture of 1,2-bis bromomethylbenzene and 1,1,2,2 tetrabromoethane.

$$C_6H_4(CH_2Br)_2 + 3Zn,\text{dust} + Br_2CH{-}CHBr_2 \longrightarrow C_{10}H_8 + 3ZnBr_2 + H_2$$

Preparation of Anthracene

Friedel-Crafts reaction of benzene with 1,1,2,2-tetrabromoethane in the presence of anhydrous aluminum chloride.

$$C_6H_6 + Br_2CH{-}CHBr_2 + C_6H_6 \xrightarrow{AlCl_3} C_{14}H_{10} + 4HBr$$

Action of zinc dust with phenyl o-tolyl ketone.

$$C_6H_5{-}CO{-}C_6H_4CH_3 + Zn,\text{dust} \xrightarrow{\text{heat}} C_{14}H_{10} + ZnO + H_2$$

Preparation of Phenanthrene

Action of sodium with 1-bromo-2-(bromomethyl) benzene.

$$BrC_6H_4CH_2Br + Na \longrightarrow \text{Phenanthrene}$$

1-bromo-2 (bromomethyl benzene) → Phenanthrene

Problem Solving Example:

Q How can we account for the fact that in the Friedel-Crafts reaction between benzene and isobutyl chloride the only product isolated is t-butyl benzene? Draw the mechanism for the reaction.

$$C_6H_6 + CH_3-\underset{}{\overset{CH_3}{\overset{|}{C}H}}-CH_2Cl \xrightarrow{AlCl_3} C_6H_5-\overset{CH_3}{\underset{CH_3}{\overset{|}{\underset{|}{C}}}}-CH_3$$

The Friedel-Crafts reaction involves the direct attachment of an alkyl group to an aromatic ring to form an alkyl benzene.

The fact that t-butyl benzene is obtained can be explained by the rearrangement of the isobutyl carbonium ion to the tertiary butyl carbonium ion.

The mechanism for the reaction is as follows:

$$CH_3-\underset{CH_3}{\underset{|}{CH}}-CH_2Cl + AlCl_3 \rightleftharpoons CH_3-\underset{CH_3}{\underset{|}{CH}}-CH_2+$$

$$CH_3-\underset{CH_3}{\underset{|}{CH}}-CH_2+ \quad \text{rearranges to} \quad CH_3-\overset{CH_3}{\overset{|}{\underset{+}{C}}}-CH_2$$

tertiary carbonium ion

via a methyl 1,2 meshift. t-butyl carbonium ion is a more stable cation. This then reacts with benzene.

$$CH_3-\overset{CH_3}{\overset{|}{\underset{+}{C}}}-CH_2 + C_6H_6 \rightleftharpoons \overset{\oplus}{C_6H_5}\begin{matrix} / \overset{CH_3}{\overset{|}{\underset{CH_3}{\underset{|}{C}}}}-CH_3 \\ \backslash H \end{matrix}$$

$$\overset{\oplus}{C_6H_5}\begin{matrix} / \overset{CH_3}{\overset{|}{\underset{CH_3}{\underset{|}{C}}}}-CH_3 \\ \backslash H \end{matrix} + AlCl_4^- \rightleftharpoons C_6H_5-\overset{CH_3}{\overset{|}{\underset{CH_3}{\underset{|}{C}}}}-CH_3 + AlCl_3$$

8.4 Properties of Benzene

A) Colorless liquid

B) Boils at 80.1°C and melts at 5.5°C

C) Symmetric benzene molecule has no net dipole moment.

D) Completely soluble in all common organic solvents and insoluble in water.

E) In general, benzene does not react or behave like alkenes.

F) Benzene undergoes mainly substitution rather than addition reactions.

G) Benzene has the tendency to retain its conjugated unsaturated ring system (highly stable to chemical reagents).

The last three properties are due to benzene's aromaticity.

8.5 Reactions of Benzene, Naphthalene, and Anthracene

Substitution Reactions of Benzene

Nitration. Generation of electrophilic $^{+}NO_2$ (nitronium ion) to attack nucleophilic benzene.

$$\underset{}{C_6H_6} + \underset{\text{Nitric acid}}{HONO_2} \xrightarrow{H_2SO_4} \underset{\text{Nitrobenzene}}{C_6H_5NO_2} + H_2O$$

Sulfonation. Generation of electrophilic SO_3 (sulfur trioxide) to attack nucleophilic benzene.

$$C_6H_6 + \underset{\text{Sulfuric acid}}{HOSO_3H} \xrightarrow{SO_3} \underset{\text{Benzenesulfonic acid}}{C_6H_5SO_3H} + H_2O$$

Halogenation. Reaction of Cl_2 or Br_2 with benzene in the presence of a Lewis-acid catalyst.

$$C_6H_6 + X_2 \xrightarrow[\text{or } FeCl_3]{AlCl_3} \underset{\text{Halobenzene}}{C_6H_5X} + HX \qquad X = Cl, Br$$

Iodination is possible through the use of special reagents, and fluorination is possible through the Balz-Schiemann reaction.

Friedel-Crafts alkylation. Reaction of benzene and alkyl halides in the presence of Lewis acids.

$$C_6H_6 + RCl \xrightarrow{AlCl_3} \underset{\text{Alkylbenzene}}{C_6H_5R} + HCl$$

Friedel-Crafts acylation. Reaction of benzene with acyl (carboxylic acid) halides in the presence of anhydrous aluminum chloride.

$$C_6H_6 + RCOCl \xrightarrow{AlCl_3} C_6H_5COR + HCl$$

Ketone

Addition Reactions of Benzene

Hydrogenation

$$C_6H_6 + 3H_2 \xrightarrow[\text{or Ni at 180°C}]{\text{Pt at R.T.}} C_6H_{12}$$

Cyclohexane

Bromination

$$C_6H_6 + 3Br_2 \xrightarrow[\text{no catalyst}]{\text{Sunlight}} C_6H_6Br_6$$

1,2,3,4,5,6-hexabromocyclohexane

Oxidation Reactions of Benzene

Vigorous Reagents

$$C_6H_6 + \text{Vig. Oxi.} \rightarrow CO_2, H_2O, HCOOH, \text{etc.}$$

Formic acid

Substitution Reactions of Naphthalene

$$C_{10}H_8 + HONO_2 \xrightarrow[H_2SO_4 \text{ conc.}]{HNO_3 \text{ conc.}} C_{10}H_7NO_2 + H_2O$$

(at 50°C)

$$C_{10}H_8 + H_2SO_4 \xrightarrow{H_2SO_4 \text{ conc.}} C_{10}H_7SO_2OH + H_2O$$

(at 80°C)

$$C_{10}H_8 + HX \xrightarrow[\text{temp}]{\text{boiling}} C_{10}H_7X + HX \quad X = Cl, Br$$

Addition Reactions of Naphthalene

$$C_{10}H_8 + 2Na + 2C_2H_5OH \xrightarrow{\text{boiling}} C_{10}H_{10}(1,4)$$

$$C_{10}H_8 + 4Na + 4C_5H_9OH \xrightarrow{\text{boiling}} C_{10}H_{12}\ (1,2,3,4)$$

$$C_{10}H_8 + 2H_2 + \text{Ni, Powder} \xrightarrow[200°C]{180°C} C_{10}H_{12}(1,2,3,4)$$

$$C_{10}H_8 + 5H_2 \xrightarrow[\text{catalyst}]{\text{heat}} C_{10}H_{18}$$

$$C_{10}H_8 + Cl_2 \xrightarrow[HCl,\,20°C]{KClO_3} C_{10}H_8Cl_2 \quad (1,4)$$

$$C_{10}H_8 + 2Cl_2 \xrightarrow[KClO_3,\,HCl,\,20°C]{} C_{10}H_8Cl_4 \quad (1,2,3,4)$$

$$C_{10}H_8 + 2Br_2 \xrightarrow{20°C} C_{10}H_8Br_4 \quad (1,2,3,4)$$

Addition Reaction of Anthracene

$$\text{Anthracene} + H_2SO_4 \longrightarrow C_{14}H_9 \cdot SO_2\ OH$$

Anthracene

1-, 2-, or 9-anthracene-sulfonic acid

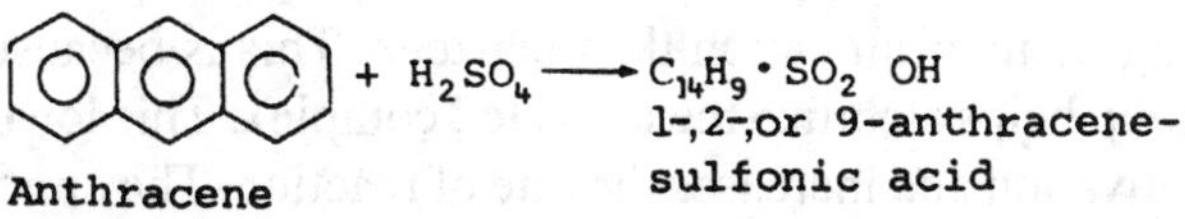

Oxidation Reaction of Anthracene

$$\text{Anthracene} + \text{air}, V_2O_5, \text{heat} \longrightarrow \text{Anthraquinone}$$

Anthracene

Anthraquinone

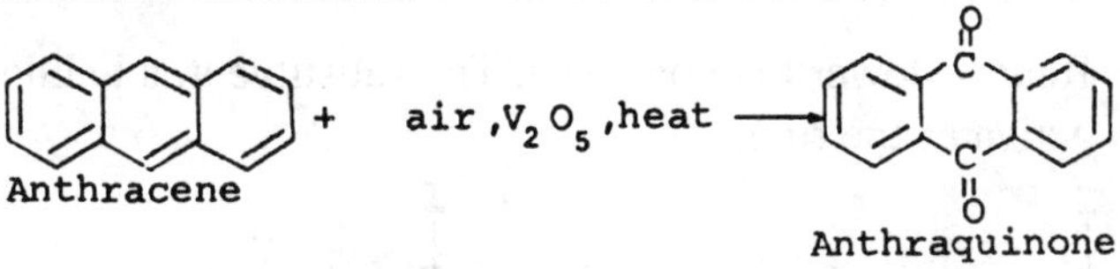

The reactions of phenanthrene resemble those of anthracene, since they are isometric.

Problem Solving Examples:

Q Give structures and names of the principal products expected from the ring monobromination of each of the following compounds. In each case, tell whether bromination will occur faster than with benzene itself.

(a) iodobenzene

(b) sec-butylbenzene

(c) acetophenone ($C_6H_5COCH_3$)

(d) phenetole ($C_6H_5OC_2H_5$)

(e) diphenylmethane ($C_6H_5CH_2C_6H_5$)

(f) benzotrifluoride ($C_6H_5CF_3$)

A Aromatic compounds with activating groups are ortho, para directors; they are electron releasing groups and will direct the electrophile to add ortho and para to the substituent. Deactivating groups are meta directors; they are electron withdrawing groups and they will direct the electrophile to add meta to the substituent. The halogens are an exception; they are deactivating groups, but they are ortho, para directors. Halogens, through their inductive effect, tend to withdraw electrons and thus to deactivate the intermediate carbonium ions. This effect is felt particularly at the ortho and para positions. Through their resonance effect, halogens release electrons. This stabilizes the carbonium ion, and it is felt most strongly at the ortho and para. The reactivity of the species is controlled by the inductive effect, whereas the resonance effect seems to govern orientation.

Those aromatic compounds with activating groups have a faster rate of ring monobromination than benzene. This is because the activating group helps stabilize the activated complex. This lowers the energy of activation and increases the rate of reaction. The opposite holds true for aromatic compounds with deactivating groups; they react slower than benzene in a ring monobromination reaction.

(a) Reacts slower than benzene. The substituent, a halide (I), deactivates the ring.

I Br Br_2/Fe + I I Br

iodobenzene o-bromoiodobenzene p-bromoiodobenzene

(b) Reacts faster than benzene. The alkyl group (isopropyl) activates the ring.

$CH_3CHCH_2CH_3$ Br_2/Fe $CH_3CHCH_2CH_3$ Br + $CH_3CHCH_2CH_3$ Br

sec-butylbenzene p-Bromosec-butylbenzene o-Bromosec-butylbenzene

(c) Reacts slower than benzene. All -COR groups are deactivating and meta-directing.

$COCH_3$ —(Br_2/Fe)→ $COCH_3$, Br

acetophenone → m-bromoacetophenone

(d) Reacts faster than benzene. –OR groups tend to be moderately activating.

OCH_2CH_3 —(Br_2/Fe)→ OCH_2CH_3, Br + OCH_2CH_3, Br

phenetole → o-bromophenetole + p-bromophenetole

(e) Reacts faster than benzene.

CH_2 —(Br_2/Fe)→ CH_2 –Br + CH_2, Br

diphenylmethane → p-bromobenzylbenzene + o-bromobenzylbenzene

(f) Reacts slower than benzene.

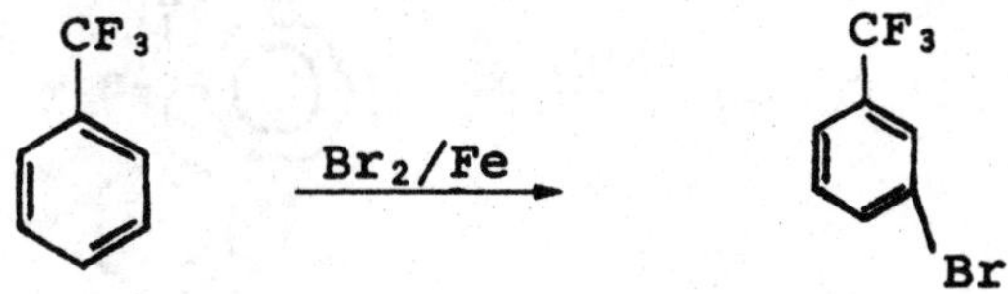

benzotrifluoride ⟶ m-bromobenzotrifluoride

Complete the following reactions. Name each organic product.

(a) Benzene + $CH_3CH(Cl)CH_3$ $\xrightarrow{AlCl_3}$

(b) CH_2Cl_2 + excess benzene $\xrightarrow{AlCl_3}$

A In alkylation reactions, the alkyl group from any alkyl halide may be substituted on the benzene ring in the presence of a Lewis acid such as aluminum chloride ($AlCl_3$).

In the first reaction, the chlorine atom is pulled off the isopropylchloride molecule and replaced by the benzene ring. This reaction is presented in Figure A. The product is isopropylbenzene.

Benzene + $CH_3CH(Cl)CH_3$ $\xrightarrow{AlCl_3}$ $C_6H_5CH(CH_3)CH_3$ +HCl

Figure A

In the second reaction, the dihalide (methylene chloride) may react with two benzene rings to yield a diphenyl compound. Since benzene is in excess, this reaction will take place. The second reaction is presented in Figure B and the product is diphenylmethane.

excess benzene + CH_2Cl_2 $\xrightarrow{AlCl_3}$ $(C_6H_5)_2CH_2$ + 2HCl

Figure B

8.6 Electrophilic Aromatic Substitution

The benzene ring may be considered an electron-rich system because of its π electrons: thus, it is a nucleophile.

The π electrons of benzene enable it to form new bonds with electron-deficient groups called electrophiles.

A disubstituted benzene ring may have one, two, or three substitution products depending on the orientation of the substituents.

The group attached to the benzene ring determines the orientation of other electrophiles and the reactivity of the benzene ring toward substitution.

A) Activating (electron-donating) groups make the substituted benzene ring more reactive than the unsubstituted benzene.

B) Deactivating (electron-withdrawing) groups make the substituted benzene ring less reactive than the unsubstituted benzene.

C) An ortho, para director is a group that causes attack to occur mainly at positions ortho and para to the group.

D) A meta director is a group that causes attack to occur mainly at position meta to the group.

Classification of Substituent Groups

Effect of Groups on Electrophilic Aromatic Substitution

Activating: Ortho, para Directors	Deactivating: Meta Directors
Strongly activating	$-NO_2$
$-NH_2$ ($-NHR$, $-NR_2$)	$-N(CH_3)_3^+$
$-OH$	$-CN$
	$-COOH(-COOR)$
Moderately activating	$-SO_3H$
$-OCH_3(-OC_2H_5$, etc.)	$-CHO$, $-COR$
$-NHCOCH_3$	
	Deactivating: Ortho, para Directors
	$-F$, $-Cl$, $-Br$, $-I$
Weakly activating	
$-C_6H_5$	
$-CH_3(-C_2H_5$, etc.)	

Ortho, para directors are activating groups, with the exception of the halogens, which are deactivating groups.

Meta directors are deactivating groups.

Rules for Predicting Orientation in Disubstituted Benzenes

A) If the groups reinforce each other, there is no problem.

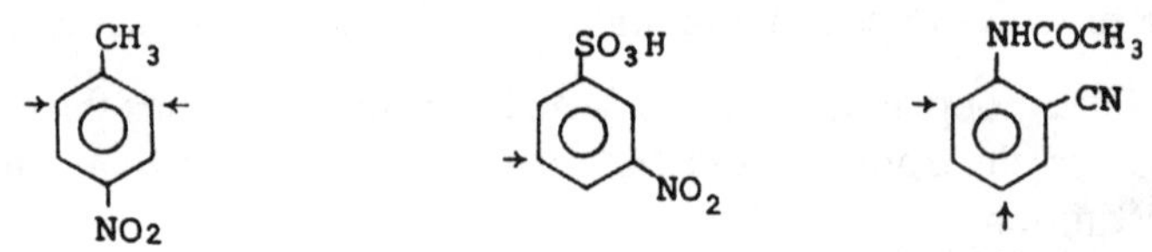

The arrows indicate the positions where electrophiles can be added.

B) If an o,p-director and m-director are not reinforcing, the o,p-director controls the orientation. The incoming group goes mainly ortho to the m-director.

Cl
NO$_2$

C) A strongly activating group, competing with a weakly activating group, controls the orientation.

OH
CH$_3$

D) When two weakly activating (deactivating) groups or two strongly activating (deactivating) groups compete, substantial amounts of both isomers are obtained; there is little preference.

Cl
CH$_3$

E) Very little substitution occurs in the sterically hindered position between two substituents.

CH$_3$
CH$_3$

Problem Solving Examples:

Explain why the $-CF_3$, $-NO_2$, and $-CHO$ groups should be meta orienting with deactivation.

A All three groups have a full or partial positive charge on the atom directly bonded to the aromatic ring. The three fluorine atoms, being highly electronegative, tend to severely decrease the electron density about the carbon atom. This gives the carbon a partial positive charge. The nitro group is represented by two principal resonance structures, both of which place a full positive charge on the nitrogen. The highly electronegative oxygen atom of the formyl group decreases the electron density about the carbon, which gives the carbon a partial positive charge. SO, $-CF_3$, $-NO_2$, and –CHO will possess a partial or full positive charge on the atom bonded to the aromatic carbon of the ring.

In such a situation, aromatic electrophilic substitution will follow a meta pathway. This is explained by the fact that when an electrophile (an electron-deficient species) attacks the aromatic ring, a carbonium ion results, that is, positive charge is introduced. If the electrophile added ortho or para to the substituent, some of this positive charge would be centered at the aromatic carbon bonded to the atom of the substituent that already possesses positive charge. This would be an unfavorable situation because of the proximity of two like charges. The situation is illustrated using the substituent $-CF_3$ and the nitronium ion ($^+NO_2$) as the electrophile below:

CF_3 + $NO_2^{\oplus}$ ⟶ (✗)

CF_3, $\frac{1}{3}$, $\frac{1}{3}+$, $+\frac{1}{3}$, H, NO_2

unfavorable

The nitronium ion added para to $-CF_3$ so that a partial positive charge now exists under the CF_3, which already has positive charge character.

Now, if the electrophile adds meta, this close proximity of like charges is avoided. The positive charge of the electrophile is not centered directly under the substituent as shown on the next page:

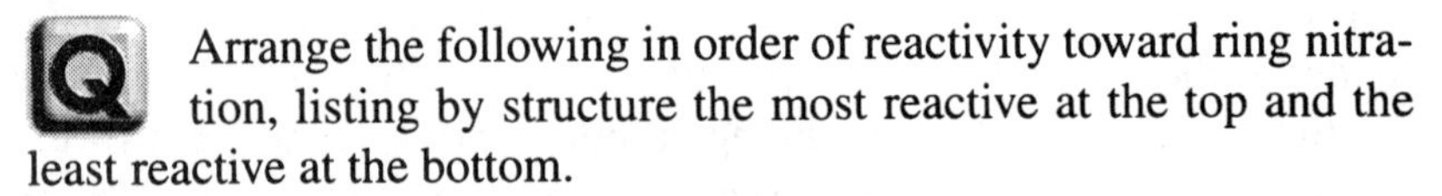

Since the close proximity of like charges is relieved, this intermediate is more stable. This, then accounts for the meta-directing orientation of $-CF_3$, $-NO_2$, and –CHO.

These three groups all withdraw electrons, so that they intensify the positive charge in the intermediate carbonium ion. This results in deactivation of the aromatic ring.

Q Arrange the following in order of reactivity toward ring nitration, listing by structure the most reactive at the top and the least reactive at the bottom.

(a) benzene, mesitylene (1,3,5–C6H3(CH3)3), toluene, m-xylene, p-xylene

(b) acetanilide ($C_6H_5NHCOCH_3$), acetophenone ($C_6H_5COCH_3$) aniline, benzene

(c) 2,4-dinitrochlorobenzene, 2,4-dinitrophenol

A Reactivity toward ring nitration depends upon the energy level of the activated complex. The higher the energy level of the activated complex, the less stable is the activated complex and the slower is the rate of reaction. Conversely, the more stable is the activated complex, the greater are the rate of reaction and reactivity.

(a) When the nitronium ion ($^+NO_2$) is added to benzene, the positive charge is distributed among three secondary carbons. In the case of ortho, para substitution in mesitylene, the positive charge in the activated complex is distributed among three tertiary carbons as shown:

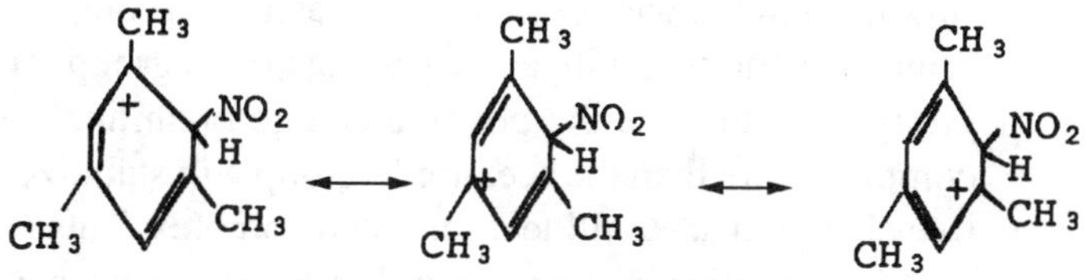

or

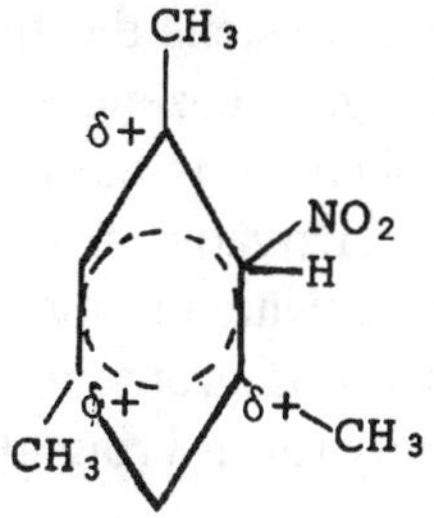

Since tertiary carbocations are more stable than secondary carbocations, the activated complex for mesitylene is more stable than for benzene. Therefore, mesitylene is more reactive toward ring nitration than is benzene. Looking at the other compounds' activated complexes for ortho, para substitution, we find: Toluene has two secondary carbocations and one tertiary carbocation; m-xylene has one secondary carbocation and two tertiary carbocations; p-xylene has two secondary carbocations and one tertiary carbocation. We see that toluene and p-xylene have the same number of secondary and tertiary carbocations in the activated complex. The activated complex for p-xylene is, however, more stable than for toluene because of the presence of an extra methyl group. The methyl group has an activating effect on the entire molecule. From the relative stabilities of the activated complexes, we can show the relative reactivities toward ring nitration to be:

mesitylene > m-xylene > p-xylene > toluene > benzene.

(b) In its activated complex, acetanilide has its positive charge distributed among three carbons, one of which is bonded to an acetamido group. The acetamido group activates the ring by stabilizing the positive charge. Acetophenone is an aromatic ring with an acetyl group. The acetyl group destabilizes the ring and will likewise destabilize the activated complex. Aniline is an aromatic ring with an activating amino group. The amino group will stabilize the positive charge in aniline's activated complex more than the acetamido group will stabilize the positive charge in acetanilide's activated complex. This is because the acetyl portion of the acetamido group is electron withdrawing and weakens the full stabilizing effect of the activating substituent. Benzene is an aromatic ring with no substituents. Acetophenone is the least reactive; it is the only compound with a deactivating group. Both aniline and acetanilide are more reactive toward nitration than benzene, because they both have an activating group, whereas benzene has none. The relative reactivity toward ring nitration is:

aniline > acetanilide > benzene > acetophenone.

(c) The activated complex for the nitration of 2,4-dinitrochlorobenzene has some positive charge nearby the deactivating chloro group. The electron withdrawing chloro group destabilizes the positive charge. In the case of 2,4-dinitrophenol, there is some positive charge nearby the activating hydroxyl group. The hydroxyl group stabilizes positive charge in the activated complex. Since both compounds have two nitro groups in the same relative positions, the reactivity toward ring nitration is:

2,4-dinitrophenol > 2,4-dinitrochlorobenzene.

8.7 Uses of Benzene, Naphthalene, and Anthracene

A) Benzene

a) Excellent solvent

b) Used in the blending of some motor fuels

c) Used in organic synthesis

B) Naphthalene

a) Used in mothballs

b) Used in fumigation of greenhouses

c) Used in the manufacture of various dyes and intermediates

d) Used as raw material in the synthesis of pthalic anhydride

e) Used as a carbon remover in motor fuels

f) Tetrahydronaphthalene is used as a solvent for fats and waxes

g) Used in the synthesis of anthraquinone

C) Anthracene

a) Used in the manufacture of dyes

b) Formerly used in the synthesis of anthraquinone

Quiz: Cyclic Hydrocarbons and Aromatic Hydrocarbons

1. What is the IUPAC name of the compound 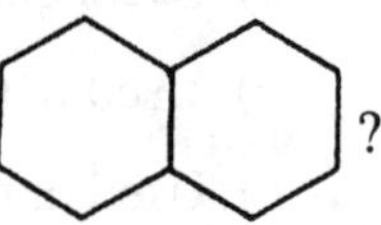 ?

   (A) (5,5,1)–bicyclodecane

   (B) (4,4,1)–bicyclodecane

   (C) (4,4,0)–bicyclodecane

   (D) (5,5,0)–bicyclodecane

   (E) (5,4,1)–bicyclodecane

2. Which of the following substituents are deactivating and ortho, para directors?

   (A) $-NH_2$

   (B) $-CH_3$

   (C) -Cl

   (D) -COOH

   (E) $-NO_2$

3. Predict the chief product in the following reaction.

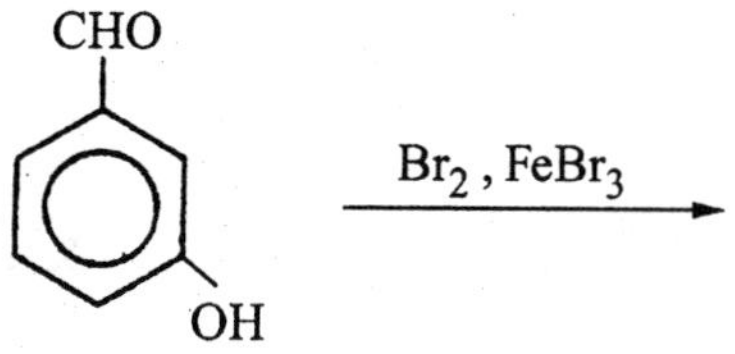

CHO, OH (meta) $\xrightarrow{Br_2,\ FeBr_3}$

(A) CHO, OH, Br

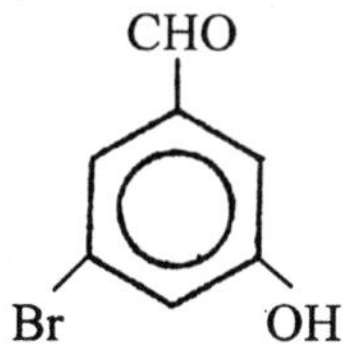

(B) CHO, Br, OH

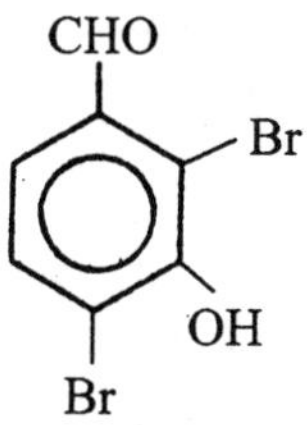

(C) Br, CHO, OH

(D) CHO, Br, OH

(E) CHO, Br, OH, Br

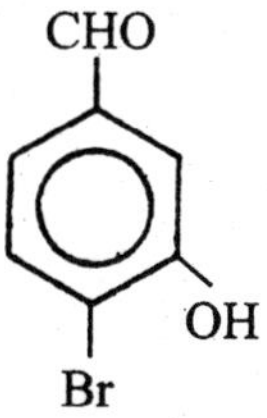

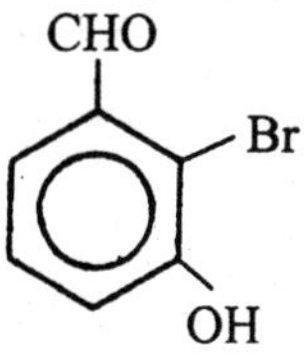

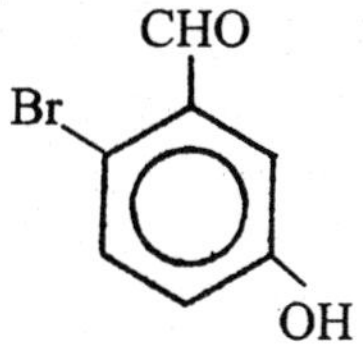

4. Which of the following electrophilic aromatic substitution(s) is (are) correct?

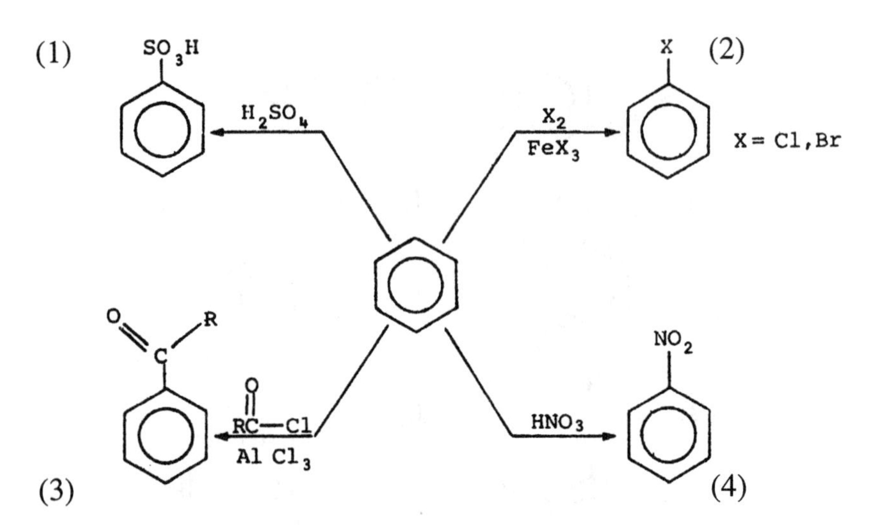

(A) (1) and (3) only.

(B) (2) and (4) only.

(C) (1), (2), and (3) only.

(D) (1), (2), and (4) only.

(E) (1), (2), (3), and (4).

5. Which one of the groups below is considered to have a deactivating effect during aromatic substitution?

(A) –OH

(B) $-NHCOCH_3$

(C) $-CH_3$

(D) $-NH_2$

(E) –CN

6. Baeyer strain is a result of

   (A) deviations from the "normal" bond angles.
   (B) deviations from a staggered arrangement.
   (C) van der Waals attractions.
   (D) twist-bond conformations.
   (E) None of the above.

7. Cyclohexane may be prepared from the reduction of a(n)

   (A) alkyne.
   (B) ketone.
   (C) benzene.
   (D) carbene.
   (E) epoxide.

8. Cyclohexenes may undergo all of the following reactions EXCEPT

   (A) cleavage.
   (B) free-radical addition.
   (C) elimination.
   (D) allylic substitution.
   (E) electrophilic addition.

9. Benzene may undergo all of the following substitution reactions EXCEPT

   (A) halogenation.
   (B) Friedel-Crafts alkylation.
   (C) Friedel-Crafts acylation.
   (D) sulfonation.
   (E) hydrogenation.

10. All of the following are true regarding naphthalene EXCEPT

    (A) used in mothballs.

    (B) may undergo oxidation reaction.

    (C) prepared by Friedel-Crafts reaction.

    (D) may undergo the Wurtz reaction.

    (E) used in the manufacture of dyes.

ANSWER KEY

1.	(C)	6.	(A)
2.	(C)	7.	(C)
3.	(C)	8.	(C)
4.	(E)	9.	(E)
5.	(E)	10.	(D)

CHAPTER 9

Aryl Halides

Aryl halides are compounds containing halogens attached directly to a benzene ring. The structural formula is ArX, where the aryl group, Ar, represents phenyl, naphthyl, etc., and their derivatives.

9.1 Nomenclature

Aryl halides are named by prefixing the name of the halide to the name of the aryl group. The terms meta, ortho, and para are used to indicate the positions of substituents on a disubstituted benzene ring. Numbers are also used to indicate the positions of the substituents on a benzene ring.

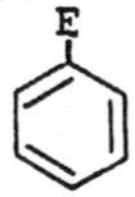

Flouro-benzene

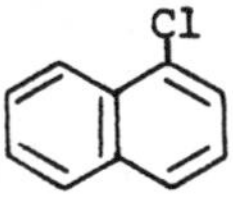

1-Chloronaphthalene

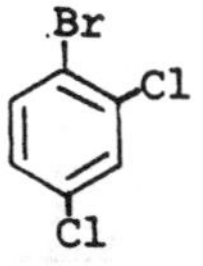

1-Bromo-2,4-Dichloro-benzene

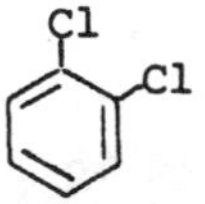

Ortho-Dichloro-benzene

9.2 Physical Properties of Aryl Halides

A) Aryl halides are colorless liquids.

B) Aryl halides are insoluble in water but soluble in organic solvents.

C) Isomeric dihalobenzenes have similar boiling points.

D) Para isomers have a melting point substantially higher than ortho and meta isomers.

E) Because of strong intracrystalline forces, the higher melting para isomer is less soluble in a given solvent than the ortho isomer. It is because of this that purification of the para isomer is often done by recrystallization.

9.3 Preparation of Aryl Halides

Halogenation by Substitution

$$ArH + X_2 \rightarrow ArX + HX \quad X_2 = Cl_2, Br_2$$

Benzene + Br_2 —(Lewis Acid, 20°C, and no light)→ Bromobenzene + HBr

Toluene + Cl_2 —(Lewis Acid, 20°C, and no light)→ o-Chloro-toluene + p-Chloro-toluene + HCl

Naphthalene + Cl_2 —(boil)→ 1-Chloronaphthalene + HCl

Anthracene + $2Cl_2$ —(100°C)→ 9,10-Dichloroanthracene + 2HCl

Halogenation by Addition

Reaction of phosphorus pentachloride with benzyl alcohol or benzaldehyde to give benzyl chloride or benzal chloride.

$$C_6H_5CH_2OH + PCl_5 \rightarrow C_6H_5CH_2Cl + POCl_3 + HCl$$

Benzyl Chloride

$$C_6H_5CHO + PCl_5 \rightarrow C_6H_5CHCl_2 + POCl_3$$

Benzal Chloride

Benzene + $3Cl_2$ $\xrightarrow[\text{bright light}]{75°C}$ 1,2,3,4,5,6-hexachlorocyclohexane

Naphthalene + $2Cl_2$ $\xrightarrow[KClO_3/HCl]{20°C}$ 1,4-dihydro naphthalene dichloride → 1,2,3,4-tetrahydro naphthalene tetrachloride

Anthracene + Br_2 $\xrightarrow{0°C}$ 9, 10 – dihydro anthracene dibromide

Sandmeyer Reaction

2-Naphthalene diazonium chloride + HCl $\xrightarrow[0-5°C]{Cu_2Cl_2}$ 2-Chloronaphthalene + N_2 + HCl

Benzene diazonium chloride + HBr $\xrightarrow[0-5°C]{Cu_2Br_2}$ Bromobenzene + N_2 + HCl

Replacement of the Nitrogen of a Diazonium Salt

$$ArH \xrightarrow[H_2SO_4]{HNO_3} ArNO_2 \xrightarrow{redn.} ArNH_2 \xrightarrow[0°]{HONO} ArN_2^+ \text{ (Diazonium salt)} \begin{cases} \xrightarrow{BF_4^-} ArF \\ \xrightarrow{CuCl} ArCl \\ \xrightarrow{CuBr} ArBr \\ \xrightarrow{I^-} ArI \end{cases} + N_2$$

Treatment of Arylthallium Compounds with Iodide

$$[ArH + Tl(OOCCF_3)_3 \longrightarrow] \; ArTl(OOCCF_3)_2 + KI \longrightarrow ArI$$

Arylthallium trifluoroacetate

For iodides only

$$\text{Toluene} \xrightarrow{Tl(OOCCF_3)_3} \xrightarrow{KI} \text{p-Iodotoluene}$$

$$\text{Benzoic acid} \xrightarrow{Tl(OOCCF_3)_3} \xrightarrow{KI} \text{o-Iodobenzoic acid}$$

Problem Solving Examples:

Q Give structures and names of the principal organic products expected from the reaction of n-propylbenzene with each of the following:

(a) Cl_2, Fe

(b) Br_2, heat, light

A N-propylbenzene is a compound made up of aromatic and alkane units. Hence, it belongs in the category of alkylbenzenes, which, in turn, are part of the group of compounds known as arenes (compounds that contain both aliphatic and aromatic units). Such compounds show two sets of chemical properties. The ring undergoes the electrophilic substitution characteristic of benzene, whereas the side chain undergoes the free-radical substitution characteristic of alkanes.

Each should modify the other. Except for oxidation and hydrogenation, these are the reactions to be expected for arenes.

(a) When ferric halogen and a halogen are added together, aromatic halogenation occurs. In this instance, it would be chlorination. The attacking species, $Cl_3\bar{F}e-\overset{+}{C}l-Cl$, does not react with the side chain of the compound; only the aromatic ring is involved. The alkyl side chain directs ortho and para, so that the following results:

$$C_6H_5CH_2CH_2CH_3 + Fe,\ Cl_2 \longrightarrow o\text{-}ClC_6H_4CH_2CH_2CH_3 + p\text{-}ClC_6H_4CH_2CH_2CH_3$$

(ortho) para

(b) It is known that halogenation of alkanes requires conditions that favor the creation of halogen radicals. High temperature or light is necessary. The halogenation of benzene involves transfer of a positive halogen (X^+) (which is protonated by acid catalysts). This means that the position of attack (whether on the benzene or on the side chain) will depend on the nature of the attacking species (an ion or a radical). Since the conditions in this problem are those that favor the production of bromine radicals, halogenation occurs exclusively to the n-propyl side chain. Hence,

$$C_6H_5CH_2CH_2CH_3 + Br_2,\ \text{heat, light} \rightarrow C_6H_5CH(Br)CH_2CH_3$$

Note: The bromine replaces a benzylic hydrogen.

$$\left[C_6H_5-\overset{|}{\underset{H}{\underset{|}{C}}}- \right]$$

i.e., the hydrogen atom attached to the carbon joined directly to the aromatic ring. These hydrogens are easy to abstract due to resonance stabilization:

CH_2 H · CH_2 H ·   H · CH_2 equivalent to CH_2 ·

The odd electron is not located on the side chain in the benzyl radical but is delocalized, being distributed about the ring.

Beginning with nitrobenzene, how would you synthesize the following compounds?

(a) Bromobenzene

(b) Benzonitrile

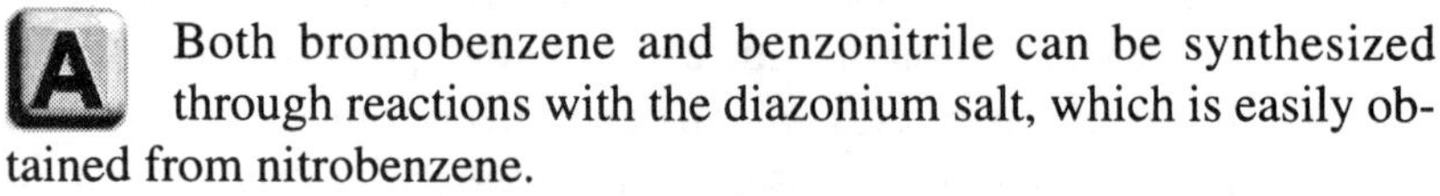

Both bromobenzene and benzonitrile can be synthesized through reactions with the diazonium salt, which is easily obtained from nitrobenzene.

$$C_6H_5NO_2 + \xrightarrow[HCl]{Sn} C_6H_5\overset{+}{N}H_3Cl^- \xrightarrow{OH^-/H_2O} C_6H_5NH_2$$

The first step in the conversion of nitrobenzene to the diazonium salt involves the formation of aniline. To aniline, HNO_2 in the presence of sulfuric acid, is added resulting in the diazonium salt.

$$C_6H_5NH_2 + HNO_2 \xrightarrow{H_2SO_4} C_6H_5\overset{+}{N}{\equiv}NCl^- \longleftrightarrow C_6H_5\ddot{N}{=}\overset{+}{\ddot{N}}Cl^-$$

The mechanism of diazonium salt formation involves the N-nitrosation of aniline.

$$H{-}\ddot{O}{-}\ddot{N}{=}\ddot{O} \xrightleftharpoons{H^+} H{-}\overset{H}{\overset{|}{O^+}}{-}N{=}O{:} \rightarrow H_2O + [N{=}O]^+$$

$$C_6H_5{-}\ddot{N}H_2 \xrightarrow{(NO)^+} C_6H_5{-}\overset{+}{N}H_2{-}N{=}O \rightleftharpoons C_6H_5{-}\overset{+}{N}H{=}N{-}\ddot{O}H$$

aniline

$$C_6H_5{-}\ddot{N}{=}N{-}\overset{+}{O}H_2 \longrightarrow C_6H_5{-}N{\equiv}N^+$$

diazonium salt

(a) Once the diazonium salt has been synthesized, bromobenzene can be obtained via the Sandmeyer reaction.

$$C_6H_5{-}N{=}N^+ + Cu_2Br_2 \longrightarrow C_6H_5Br + N_2\uparrow$$

bromobenzene

(b) The Sandmeyer reaction is also useful in obtaining cyano compounds.

$$C_6H_5{-}\overset{+}{N}{=}N + Cu_2(CN)_2 \longrightarrow C_6H_5{-}C{\equiv}N + N_2\uparrow$$

9.4 Reaction of Aryl Halides

Formation of Grignard Reagent

$$ArBr + Mg \xrightarrow{\text{dry ether}} ArMgBr$$

$$ArCl + Mg \xrightarrow[\text{furan}]{\text{tetrahydro-}} ArMgCl$$

Substitution in the ring (electrophilic aromatic substitution). X: Deactivates and directs ortho, para in electrophilic aromatic substitution.

Nucleophilic aromatic substitution (bimolecular displacement).

$$Ar{:}X + {:}B^- \rightarrow Ar{:}B + {:}X^-$$

Ar must contain strongly electron-withdrawing groups ortho and/or para to it.

2,4-Dinitrochlorobenzene + NaOH → (sodium 2,4-dinitrophenoxide, O^-Na^+) $\xrightarrow{H^+}$ 2,4-Dinitrophenol

2,4-Dinitrochlorobenzene + NH_3 → 2,4-Dinitroaniline

2,4-Dinitrochlorobenzene + $NaOC_2H_5$ → 2,4-Dinitrophenylethyl ether

Replacement of the Halogen Atom

$$C_6H_5X + HNH_2 \text{ (excess, aq.)} \xrightarrow[\text{pressure}]{\text{heat}} C_6H_5NH_2 + HX$$

aniline

$$C_6H_5X + NaOH \text{ aq.} \xrightarrow[\text{pressure}]{\text{heat}} C_6H_5OH + NaX$$

phenol

$$C_6H_5X + 2Na + RX \rightarrow C_6H_5R + 2NaX$$

$$C_6H_5X + XMgR \rightarrow C_6H_5R + MgX_2$$

Replacement of the Hydrogen Atom

$$X - C_6H_4H + X \cdot X \xrightarrow[\text{catalyst}]{\text{diffused light, 20°C}} X - C_6H_4 - X + HX$$

$$X - C_6H_4H + HONO_2 \xrightarrow[H_2SO_4\text{, Conc. } 50\text{-}60°C]{HNO_3\text{, Conc.}} X - C_6H_4 - NO_2 + H_2O$$

$$X - C_6H_4 - H + HOSO_2OH \xrightarrow{H_2SO_4/SO_3} X - C_6H_4 - SO_2OH + H_2O$$

Nucleophilic aromatic substitution. Elimination-addition (benzyne) mechanism:

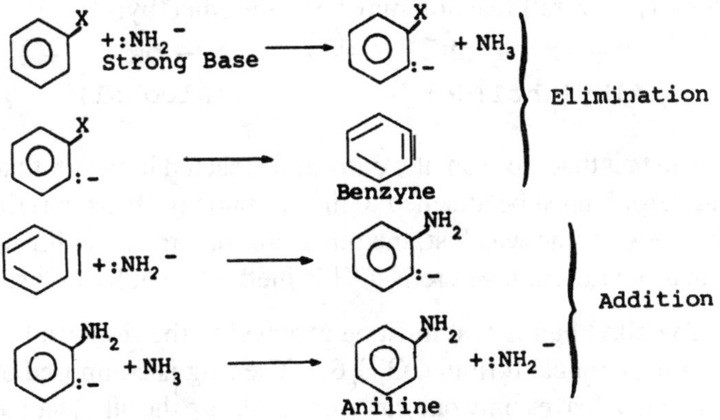

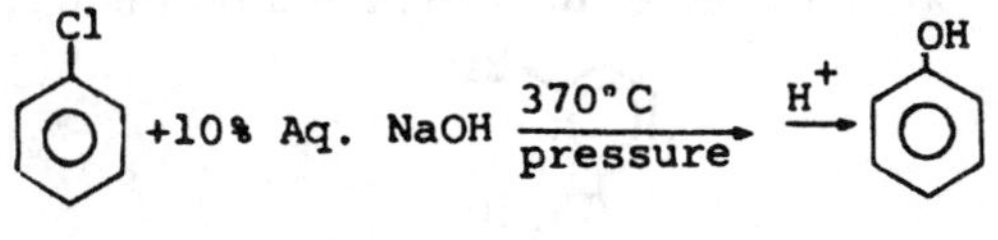

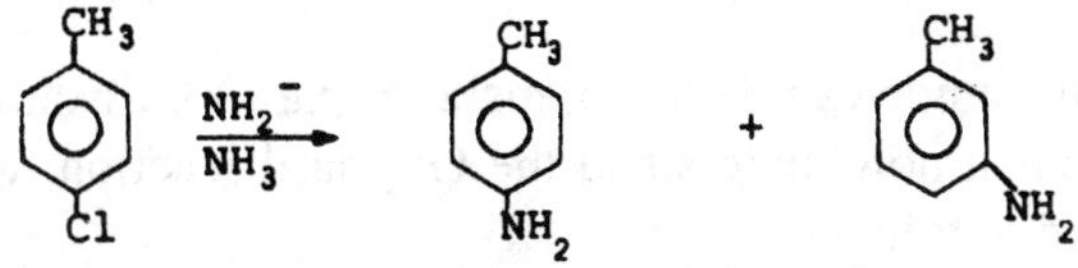

X + :NH_2^- Strong Base → X :– + NH_3

X :– → Benzyne

Elimination

Benzyne + :NH_2^- → NH_2 :–

NH_2 :– + NH_3 → Aniline + :NH_2^-

Addition

Cl + 10% Aq. NaOH $\xrightarrow[\text{pressure}]{370°C}$ $\xrightarrow{H^+}$ OH

CH_3 / Cl $\xrightarrow[NH_3]{NH_2^-}$ CH_3 / NH_2 + CH_3 / NH_2

p-Chlorotoluene p-and m-Aminotoluene

Problem Solving Examples:

Q An aromatic dibromide $C_7H_6Br_2$ reacted with aqueous sodium hydroxide. The product of this reaction had lost only one bromo group to give the product C_7H_7BrO. When the dibromide was converted to a Grignard reagent and then hydrolyzed, the product was toluene. Determine the structure of the dibromide.

A It is known that alkyl halides can be converted to alcohols by hydrolysis using a nucleophilic displacement reaction. (Note: This reaction is NOT the same as dehydrohalogenation where the sodium hydroxide is in alcohol. In hydrolysis, the sodium hydroxide is in water.) The overall reaction may be represented by:

$$\underset{\text{(Alkyl halide)}}{R—X} + OH^- \xrightarrow{H_2O} \underset{\text{(Alcohol)}}{R—OH} + X^-$$

The fact that the aromatic dibromide reacted indicates that an alkyl halide group must be attached to the aromatic (or benzene) ring. Since only one bromine was lost, the other bromine must be directly bound to the aromatic ring, where NaOH is ineffective in hydrolysis.

The alkyl halide that must be attached to the ring can be deduced from the chemical formula, $C_7H_6Br_2$. The ring is composed of six carbons, which leaves only one carbon to make up the alkyl halide. Hence, the alkyl halide attached must be CH_2Br. Recalling that the other bromine must be directly attached to the ring, the dibromide can now be written:

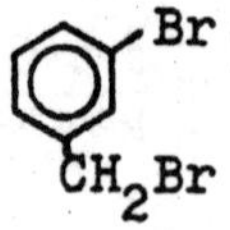

Br

CH_2Br

(Note: The dibromide can be ortho, meta, or para.) By examining this structure, one can now understand the Grignard reaction, which will ultimately lead to toluene.

Alkyl halides and aryl halides can react with magnesium in anhydrous ether to give alkylmagnesium halides, RMgX, which are known as Grignard reagents. The Grignard reagents can then be hydrolyzed by water or acid to the corresponding hydrocarbon as shown:

$$R—MgX + HOH \rightleftharpoons R—H + MgXOH$$

(Hydrogen-carbon)

The sequence that generated toluene can now be written:

$$C_6H_4(Br)(CH_2Br) + 2\,Mg \rightarrow C_6H_4(MgBr)(CH_2MgBr) \xrightarrow{2\,HOH} C_6H_5CH_3 + 2MgBrOH$$

(Toluene)

Synthesize a ketone from the following precursor using the reaction of acid chlorides with organocadmium compounds.

(a) $CH_3CH_2CH_2CH_2MgBr$

Butyl Magnesium Bromide

(b) m-Bromotoluene ($CH_3C_6H_4Br$)

A Grignard reagents react with dry cadmium chloride to yield the corresponding organocadmium compounds, which react with acid chlorides to yield ketones. To make this concept clearer, the general equation for the synthesis is:

$$2R'MgX + CdCl_2 \rightarrow R'_2Cd + 2\,MgXCl \quad \text{Balanced}$$

Grignard reagent

$$R'_2Cd + 2RCCl(=O) \rightarrow 2R\text{-}C(=O)\text{-}R' + CdCl_2$$

ketone

In this type of synthesis R' must be either aryl or primary alkyl.

Only organocadmium compounds containing aryl or primary alkyl groups are stable enough for use. In spite of this limitation, the method is one of the most valuable for the synthesis of ketones.

In this ketone synthesis, (1) the Grignard reagent reacts with cadmium chloride to yield an organocadmium compound, plus a magnesium dihalide. (2) The organocadmium is then reacted with an acid chloride to yield a ketone plus cadmium chloride.

Solving the problem to yield a ketone:

(a) $2\,CH_3CH_2CH_2CH_2MgBr \xrightarrow{CdCl_2} (CH_3CH_2CH_2CH_2)_2Cd + 2MgBrCl$

Butyl magnesium bromide → Di-n-butylcadmium

Di-n-butylcadmium + $2CH_3CH(CH_3)C(=O)Cl$ (Isobutyryl chloride) → $2\,CH_3CH_2CH_2CH_2C(=O)CH(CH_3)CH_3$

n-butyl isopropyl ketone
(2-methyl-3-heptanone)

(b) m-bromotoluene ($CH_3C_6H_4Br$) $\xrightarrow[\text{ether}]{Mg}$ $CH_3C_6H_4MgBr$ $\xrightarrow{CdCl_2}$ $(CH_3C_6H_4)_2Cd$

$(CH_3C_6H_4)_2Cd$ + $2CH_3CH_2CH_2C(=O)Cl$ (butyryl chloride) → $CH_3C_6H_4C(=O)CH_2CH_2CH_3$

n-propyl m-tolyl ketone

Ethers and Epoxides

Ethers are hydrocarbon derivatives in which two alkyl or aryl groups are attached to an oxygen atom. The structural formula of an ether is R–O–R', where R and R' may or may not be the same.

10.1 Structures of Ethers

Ethers and alcohols are metameric. They are functional isomers of alcohols with the same elemental composition.

$$CH_3OCH_3 \text{ and } CH_3CH_2OH$$

Problem Solving Example:

Write structural formulas for:

(a) methyl ether

(b) isopropyl methyl ether

(c) 3-methoxyhexane

(d) 1,2-epoxypentane

A The characteristic functional group in ethers is the oxygen atom single bonded to two carbon atoms. The general formulas for these compounds are R'OR, ROAr, and Ar'OAr, where R and R' are either different or identical alkyl groups and Ar and Ar' represent different or identical aryl groups.

When naming simple ethers, both of the attached groups are named in alphabetical order followed by the word "ether." For example,

$$CH_3 - O - CH_2CH_3$$

is called ethyl methyl ether.

(a) If both of the attached groups are the same, the compound is a symmetrical ether, and only one group need be named. For example, the compound $CH_3CH_2 - O - CH_2CH_3$ is usually called ethyl ether. Often to avoid confusion this compound is referred to as diethyl ether. The structure for methyl ether is analogous to this. There is one methyl group on each side of the oxygen. The structure can be written as follows:

$$CH_3 - O - CH_3.$$

This compound is also called dimethyl ether.

(b) Isopropyl methyl ether. This is a simple asymmetrical ether. From its name, one knows that there is a methyl group on one side of the oxygen and an isopropyl group on the other. The structure for this compound is written:

```
CH3-O-CH-CH3
      |
      CH3
```

(c) In more complicated ethers, the ether grouping, R – O – (an alkoxy group), may be named as a substituent on a longer chain. A methoxy group is written $- O - CH_3$. In 3-methoxy-hexane, the methoxy group is attached to the carbon at the 3 position of the hexane chain.

```
CH3-CH2-CH-CH2-CH2-CH3
        |
        O-CH3
```

3-methoxyhexane

(d) Epoxides are a class of cyclic ethers in which the ether oxygen is included in a three-membered ring,

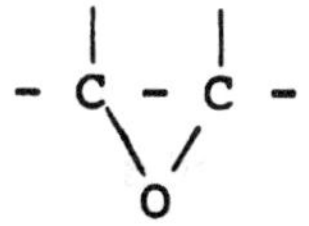

```
  |     |
- C  -  C -
   \   /
     O
```

Ethylene oxide is drawn as shown:

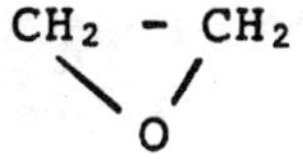

```
CH2 - CH2
   \  /
    O
```

The epoxides are named by numbering the alkyl chain and indicating the two carbons to which the oxygen is attached. The prefix "epoxy" denotes this functional group. Ethylene oxide can also be called epoxyethane. In 1,2-epoxypentane, the oxygen is bound to the first two carbons of the pentane chain.

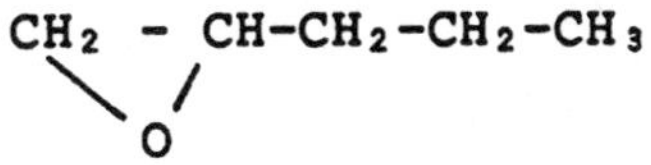

```
CH2 - CH-CH2-CH2-CH3
   \  /
    O
```

1,2-epoxypentane

10.2 Nomenclature (IUPAC System)

Common Names

The attached groups are named in alphabetical order, followed by the word ether.

CH_3CH_2-O-$CH_2CH_2CH_3$
Ethyl propyl ether

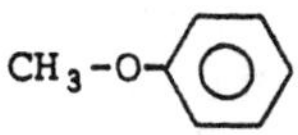

CH_3-O-C_6H_5 (benzene ring)
Methyl phenyl ether

For symmetrical ethers (having the same groups), the compound is named using either the name of the group or the prefix "di–."

Example $CH_3 - O - CH_3$

Methyl ether or
Dimethyl ether

In the IUPAC system, ethers are named ao· lk oxyalkanes. The larger alkyl group is chosen as the stem.

Example

```
          Cl
          |
    CH3-C  —  CH—CH3
          |    |
          Cl   OCH2CH3
```

3,3-dichloro-2-ethoxybutane

Problem Solving Example:

Name each of the following compounds:

(a) $CH_3 - O - CH_2 - CH(CH_3)_2$

(b) $(CH_3CH_2CH_2)_2 - O$

(c) $(CH_3)_2CH - O - CH_2CH_2CH_2CH_3$

(d) $CH_3 - O - CH_2CH_2 - O - CH_3$

(e) $CH_3CH_2 - O - CH_2CH_2CH_2CH_2CH_2 - OH$

(f)
```
        Cl
        |
    CH3CHCH2CH-CH2
             \ /
              O
```

A Ethers have the general formula of R-O-R', where R and R' may represent an alkyl or aryl group. To name many ethers, one usually names the two groups that are attached to oxygen, and follows these names with the word "ether."

(a) The compound

```
                CH3
               /
    CH3-O-CH2CH
               \
                CH3
```

has the oxygen of the ether linkage bonded to methyl and isobutyl groups. It is called isobutyl methyl ether.

(b) $CH_3CH_2CH_2 - O - CH_2CH_2CH_3$ is a symmetrical molecule because the two groups bound to the oxygen are identical. In this compound, the two groups are n-propyls. Thus, it is named dipropyl ether, or n-propyl ether. It is understood that if the name only contains one substituent group, the compound is symmetrical.

(c) The compound

```
CH3
   \
    CH-O-CH2CH2CH2CH3
   /
CH3
```

contains an isopropyl group and an n-butyl group. It is called n-butyl isopropyl ether.

(d) When an ether contains groups that do not have simple names, the compound may be named as an alkoxy derivative. The alkoxy functional group can be written as RO, where R may be an alkyl group. Since $CH_3 - O - CH_2CH_2 - O - CH_3$ contains two ether functional groups, which make naming very complicated, one may name the compound as a derivative of ethane. The two substituent groups are both methoxy (CH_3O–) at carbons 1 and 2, respectively. Thus, the compound can be named as 1,2-dimethoxy-ethane.

(e) $CH_3CH_2 - O - CH_2CH_2CH_2CH_2CH_2OH$ has two functional groups, alkoxy (or ether) and hydroxy (or alcohol). In the IUPAC system, the alcohol functional group takes precedence over the ether in naming. (Exception can be made if the naming is overly complicated when named as an alcohol.)

The compound in question can be named as an alcohol with an alkoxy substitution. The hydrocarbon containing the hydroxyl has five carbons. Thus, the compound is a derivative of n-pentanol. The alkoxy substitution is an ethoxy (CH_3CH_2O –) group at the number 5 carbon (the hydroxyl group is attached to the first carbon). The complete name for the compound is 5-ethoxy-l-pentanol.

(f) The compound

```
        Cl
        |
CH3CCHCH2CH-CH2
           \ /
            O
```

contains a three-member ring that includes an oxygen atom. Such compounds are called epoxides. The epoxides are named by numbering the alkyl chain and indicating the two carbons to which the oxygen is attached. The prefix "epoxy" denotes this functional group. Here, the alkyl group contains five carbons; it is a pentane. The epoxide oxygen is attached to carbons 1 and 2, so that the name becomes 1,2-epoxypentane. The compound also has a chlorine substitution at carbon number 4. The full name for

```
      Cl
      |
CH3CHCH2-CH-CH2
          \ /
           O
```

is 4-chloro-l,2-epoxypentane.

10.3 Physical and Chemical Properties of Ethers

A) Ethers are much more volatile than isomeric alcohols.

B) Ethyl ether is colorless, highly volatile, flammable, less dense than and only partially soluble in water; it has a characteristic odor, and is an excellent solvent.

C) Ethers are fairly unreactive to many reagents.

D) Tetrahydrofuran (THF), a cyclic ether, has a boiling point of 67°C, is an important solvent, and is miscible with water.

E) Ethers undergo cleavage in the presence of strong acids.

F) Volatility, flammability, and solubility in water decreases with an increase in C-content. The densities show a gradual increase with increasing molecular weight.

10.4 Preparation of Ethers

Williamson Reaction

$RX + NaOR' \rightarrow ROR' + NaX$ R' = alkyl or aryl alkoxide

Example $CH_3CH_2I + NaOCH_2CH_3$ Æ $CH_3CH_2OCH_2CH_3 + NaI$

ethyl iodide, sodium ethoxide, ethyl ether

Reaction of an Alkyl Halide Silver Oxide

Example

$$RX + Ag_2O + XR' \xrightarrow{\text{heat}} ROR' + 2AgX$$

$$2CH_3I + Ag_2O \xrightarrow{\text{heat}} CH_3OCH_3 + 2AgI$$

methyl iodide, methyl ether

Dehydration of Alcohols

a) $ROH + HOSO_2OH \xrightarrow{\text{cold}} ROSO_2OH + H_2O$

$ROSO_2OH + HOR' \xrightarrow{\text{heat}} ROR' + H_2SO_4$

Example

$$CH_3CH_2OH + HOSO_2OH \xrightarrow{\text{cold}} CH_3CH_2OSO_2OH + H_2O$$

ethanol, ethyl hydrogen sulfate

$$CH_3CH_2OSO_2OH + HOCH_2CH_3 \xrightarrow{\text{heat}} CH_3CH_2OCH_2CH_3 + H_2SO_4$$

ethyl ether

b) $ROH \xrightarrow[240\text{–}260°C]{Al_2O_3} ROR + H_2O$

Example

$$2CH_3CH_2OH \xrightarrow[240-250°C]{Al_2O_3} CH_3CH_2OCH_2CH_3 + H_2O$$

ethyl alcohol → ethyl ether

Alkoxymercuration–Demercuration

Example

alkoxymercuration demercuration

$$\rangle C{=}C\langle + ROH + Hg(OOCCF_3)_2 \rightarrow -\underset{OR}{\overset{|}{\underset{|}{C}}}-\underset{HgOOCCF_3}{\overset{|}{\underset{|}{C}}}- \xrightarrow{NaBH_4} -\underset{OR}{\overset{|}{\underset{|}{C}}}-\underset{H}{\overset{|}{\underset{|}{C}}}-$$

ether

Example

$$CH_3-\underset{CH_3}{\overset{CH_3}{\underset{|}{\overset{|}{C}}}}-CH{=}CH_2 + CH_3CH_2OH \xrightarrow{Hg(OOCCF_3)_2} \xrightarrow{NaBH_4}$$

3,3-dimethyl-1-butene

$$CH_3-\underset{CH_3}{\overset{CH_3}{\underset{|}{\overset{|}{C}}}}-\underset{OC_2H_5}{\underset{|}{CH}}-CH_3$$

3,3-dimethyl-2-ethoxybutane

Problem Solving Examples:

Starting with any alcohols, outline all steps in the synthesis of n-hexyl isopropyl ether, using the Williamson method.

A The Williamson synthesis of ethers is important because of its versatility in the laboratory. This method of synthesis can be used to make asymmetrical ethers as well as symmetrical ethers. In the Williamson synthesis, an alkyl halide is allowed to react with a sodium alkoxide (or sodium phenoxide).

$$R-X + R'-O^-Na^+ \rightarrow R\text{–}O\text{–}R' + Na^+X^-,$$

Alkyl halide	Sodium alkoxide		Ether

where R represents an alkyl group and X a halide (Cl, Br, I, or F). [The yield from RX is $CH_3 > 1° > 2° (>3°)$.] The sodium alkoxide is made by direct action of sodium metal on dry alcohols:

$$R - OH + Na \rightarrow R - O^-Na^+ + H_2 \uparrow$$

Sodium alkoxide

As an example,

$$CH_3 - CH_2OH + Na \rightarrow CH_3 - CH_2 - O^-Na^+$$

Ethanol   Sodium ethoxide

Alkyl halides can be prepared from alcohols by use of phosphorus trihalides:

$$R - OH + PX_3 \rightarrow RX + H_3PO_3$$

Alkyl halide

As an example,

```
     CH3                          CH3
     |                            |
CH3-CH -CH2OH + PCl3         CH3-CH-CH2Cl         +  H3PO3
```

Isobutyl alcohol   Isobutyl chloride

The Williamson synthesis involves nucleophilic substitution of the alkoxide ion for the halide ion.

The compound n-hexyl isopropyl ether is an asymmetrical ether with the formula

```
                   CH3
                   |
CH3(CH2)4CH2-O-CH-CH3
```

and can be prepared from the corresponding alcohols, n-hexyl alcohol and isopropyl alcohol. The complete reactions are:

$$CH_3\text{-}\overset{\overset{CH_3}{|}}{C}H\text{-}OH + Na \longrightarrow CH_3\text{-}\overset{\overset{CH_3}{|}}{C}H\text{-}O^-Na^+$$

Isopropyl alcohol — Sodium isopropoxide

$$CH_3(CH_2)_4CH_2OH + PBr_3 \longrightarrow CH_3(CH_2)_4CH_2Br + H_3PO_3$$

n-Hexyl alcohol — n-Hexyl bromide

$$CH_3(CH_2)_4CH_2Br + CH_3\text{-}\overset{\overset{CH_3}{|}}{C}H\text{-}O^-Na^+ \rightarrow CH_3(CH_2)_4CH_2\text{-}O\text{-}\underset{\underset{CH_3}{|}}{C}H\text{-}CH_3 + NaBr$$

n-Hexyl bromide + isopropoxide — n-Hexyl isopropyl ether

Q In ether formation by dehydration, as in most other cases of nucleophilic substitution, there is a competing elimination reaction. What is this reaction and what products does it yield? For what alcohols would elimination be most important?

A Ethers have the general formula R – O – R', where R and R' can be aryl and/or alkyl. Ethers can be formed by two alcohol molecules in the presence of strong acid to release a water molecule. This is called acid-catalyzed dehydration of alcohols to form ethers.

The mechanism of this reaction resembles that of substitution. It involves the protonation of one alcohol and release of a water molecule to form a carbonium ion. A carbonium ion is a strong electrophile. It attacks the electron-rich oxygen of another alcohol, which releases a proton to form an ether.

$$R\text{-}O\text{-}H \xrightarrow[\text{heat}]{H_2SO_4} R^+ + H_2O$$

$$R^+ + :\ddot{O}(H)\text{-}R \longrightarrow R\text{-}O\text{-}R + H^+$$

Under closer scrutiny, one can see that in the process of strong acid protonation of alcohol, another reaction can occur instead of the substitution attack on another alcohol. A carbonium ion is a highly reactive species. It can participate in a substitution reaction, as in ether formation, or undergo an elimination reaction to form an alkene. The latter reaction is most prominent in tertiary alcohols.

A tertiary alcohol has the general formula of

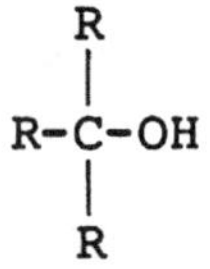

```
    R
    |
  R-C-OH
    |
    R
```

It is a bulky molecule. This bulkiness makes the positively charged carbon of the carbonium ion formed less accessible. Therefore, the tertiary carbonium ion can release a proton (H+) to obtain the alkene, which will be stabilized by the alkyl substituents.

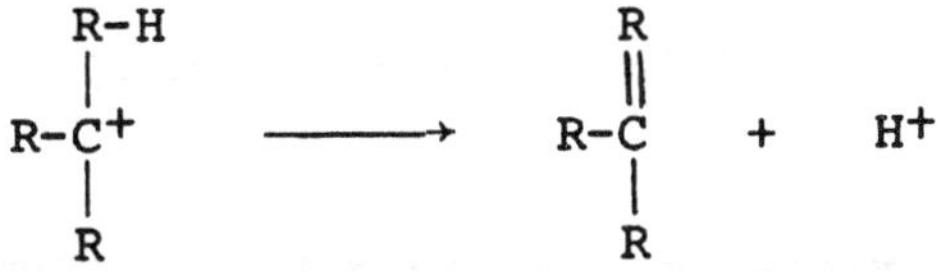

```
   R-H              R
   |                ||
 R-C+   ------>   R-C     +    H+
   |                |
   R                R
```

The dehydration of alcohols to ethers rather than alkenes also reflects the choice of reaction conditions.

10.5 Reactions of Ethers

Single Cleavage at the Oxygen Linkage

a) HI, cold and conc.

$$R\text{-}\boxed{O\text{-}R' + H}I \rightarrow R'\text{-}OH + R - I$$

b) Sulfuric acid, conc., heat

$$R\ \boxed{O\text{-}R' + H}O\text{-}SO_2\text{-}OH \rightarrow \dot{R}\text{-}OH + R\text{-}O\text{-}SO_2\text{-}OH$$

c) Steam under pressure, 150°C

$$R\text{-}\boxed{O\text{-}R' + H}OH \rightarrow R\text{-}OH + R'\text{-}OH$$

Double Cleavage

a) Phosphorus pentachloride, heat

$$R - O - R + PCl_5 \longrightarrow 2R - Cl + POCl_3$$

b) HI, conc., heat

$$R-O-R+2HI \longrightarrow 2R-I+H_2O$$

c) Sulfuric acid, heat

$$R\text{-}O\text{-}R + 2HO\text{-}SO_2\text{-}OH \xrightarrow{heat} H_2O + 2R\text{-}O\text{-}SO_2\text{-}OH$$

Substitution on the Hydrocarbon Chain

Example

$$\underset{\substack{|\\R\text{-}O}}{R\text{-}HC}\boxed{H + X}\text{-}X \rightarrow \underset{\substack{|\\R\text{-}O}}{R\text{-}HC}\text{-}X + HX \quad X = Cl, Br$$

$$CH_3\text{-}CH_2\text{-}O\text{-}CH_2\text{-}CH_3 + Cl_2 \xrightarrow{dark} CH_3CH\,Cl\text{-}O\text{-}CH_2CH_3 + HCl$$

$$CH_3\text{-}CH_2\text{-}O\text{-}CH_2\text{-}CH_3 + 10Cl_2 \xrightarrow{light} CCl_3\text{-}CCl_2\text{-}O\text{-}CCl_2\text{-}CCl_3 + 10HCl$$

Problem Solving Examples:

Q (a) Upon treatment with sulfuric acid, a mixture of ethyl and n-propyl alcohols yields a mixture of three ethers. What are they? (b) On the other hand, a mixture of tertbutyl alcohol and ethyl alcohol gives a good yield of a single ether. What ether is it likely to be? How can the good yield be accounted for?

(a) Ethers can be synthesized by the acid-catalyzed dehydration of alcohols. However, this reaction mechanism is only useful for the synthesis of symmetrical ethers (ROR, where both alkyl or aryl groups are identical). It is inefficient in the synthesis of unsymmetrical ethers (where the alkyl and/or aryl groups are not identical) because it gives a mixture of products. This can be explained by the reaction mechanism.

The reaction in question is an acid-catalyzed dehydration. The first step of this reaction is the protonation of alcohol by the acid, H_2SO_4. Since the reaction mixture contains two different alcohols, each one of them can be protonated. This results in two different kinds of carbonium ions after release of water molecules. Each of these two carbonium ions can attack either alcohol to form an ether. If the ethyl cation ($CH_3CH_2^+$) reacts with ethyl alcohol, diethyl ether is formed. If the n-propyl cation ($CH_3CH_2CH_2^+$) reacts with n-propyl alcohol, di-n-propyl ether is formed. And if ethyl cation reacts with n-propyl alcohol, or n-propyl cation reacts with ethyl alcohol, ethyl n-propyl ether is formed. Since ethyl and n-propyl cations are formed at the same rate, the product will contain a mixture of diethyl ether, di-n-propyl ether, and ethyl n-propyl ether.

(b) In a mixture of tert-butyl alcohol and ethyl alcohol, the major product is ethyl t-butyl ether. This product is in good yield due to the kinetic effect.

As mentioned earlier, the first step of an acid-catalyzed dehydration of alcohol is the protonation of the alcohol. The elimination of a water molecule to form the carbonium ion from the protonated alcohol is the slowest step. Therefore, it is the rate-limiting step. The rate at which a carbonium ion forms is dependent on its stability. It can be formed fastest if there are electron-releasing groups to stabilize the positive charge. The rate of carbonium ion formation is 3° > 2° > 1° methyl. This is also the order of stability of carbonium ions. Since tert-butyl alcohol can form a 3° cation, whereas ethyl alcohol can only form a 1° carbonium ion, tert-butyl-cation

$$\left[CH_3-\overset{\displaystyle CH_3}{\underset{\displaystyle CH_3}{\overset{|}{\underset{|}{C^+}}}} \right]$$

should form the fastest and predominate over ethyl cation ($CH_3CH_2^+$) in the mixture.

Since the tert-butyl group is bulky, the 3° carbonium ion formed tends to attack ethyl alcohol more often than tert-butyl alcohol in the subsequent substitution reaction. The reactions are shown on the next page:

$$CH_3-\underset{CH_3}{\overset{CH_3}{\overset{|}{\underset{|}{C}}}}-OH \xrightarrow[-H_2O]{H^+} CH_3-\underset{CH_3}{\overset{CH_3}{\overset{|}{\underset{|}{C^+}}}} \xrightarrow{CH_3CH_2OH} CH_3-\underset{CH_3}{\overset{CH_3}{\overset{|}{\underset{|}{C}}}}-\overset{H}{\overset{|}{O^+}}-CH_2CH_3$$

$$\xrightarrow{-H^+} CH_3-\underset{CH_3}{\overset{CH_3}{\overset{|}{\underset{|}{C}}}}-O-CH_2CH_3$$

ethyl t-butyl ether

In conclusion, the ethyl t-butyl ether should be the predominant ether because of the speed in which t-butyl cation forms and the difficulty of this cation attacking the t-butyl alcohol for steric reasons.

Q Insofar as the ether linkage is concerned, ethers undergo just one type of reaction, cleavage by acids. Discuss this reaction in terms of the mechanism, conditions, and products obtained.

A While the ether linkage (R – O – R') is stable toward bases, reducing agents, and oxidizing agents, they can be cleaved by acids as illustrated:

$$R-O-R' + HX \rightarrow R-X + R'OH \xrightarrow{HX} R'X$$

$$Ar-O-R + HX \rightarrow R-X + Ar-OH$$

This reaction will occur only under vigorous conditions (that is, concentrated acids at high temperatures). The alkyl ether reacts with the acid to yield an alkyl halide and an alcohol. The alcohol may continue to react with acid to generate a second mole of alkyl halide. An aryl alkyl ether yields a phenol and an alkyl halide because cleavage occurs at the alkyl-oxygen bond, and not the aryl-oxygen due to the latter's low reactivity. For example,

$$C_6H_5-OCH_3 \xrightarrow[120-130°]{57\%\ HI} C_6H_5-OH + CH_3I$$

Anisole — Phenol — Methyl iodide

Selection of the acid can be important because of their different reactivities as indicated:

Reactivity of HX: HI > HBr > HCl.

The cleavage of the ether by acid involves nucleophilic attack by halide ion on the protonated ether. This causes displacement of the weakly basic alcohol molecule as illustrated:

$$R\ddot{O}R' + HX \rightleftharpoons R\overset{+}{\underset{\cdot\cdot}{O}}(H)R' + X^- \xrightarrow[S_N2]{S_N1 \text{ or}} RX + R'OH$$

(Protonated ether) (Good leaving group)

As mentioned, the reaction may proceed by substitution nucleophilic unimolecular (S_N1) or substitution nucleophilic bimolecular (S_N2). The conditions and structure of the ether will determine which way the reaction goes. S_N1 and S_N2 are outlined below.

S_N1

(1) $R\overset{+}{O}(H)R' \xrightarrow{\text{slow}} R^+ + HOR'$

(2) $R^+ + X^- \xrightarrow{\text{fast}} R\text{-}X$

S_N2

$$R\overset{+}{O}(H)R' + X^- \longrightarrow \left[\overset{\delta-}{X}\text{---}R\text{---}\underset{\delta+}{O}(H)R'\right] \longrightarrow RX + HOR'$$

Since the S_N1 mechanism proceeds through an intermediate carbonium ion, it might be expected that a tertiary alkyl group undergoes S_N1 displacement, whereas a primary alkyl group undergoes S_N2 displacement.

10.6 Structure of Epoxides

Epoxides are cyclic ethers in which the oxygen is included in a three-membered ring.

$$\underset{CH_2-CH_2}{\overset{O}{\wedge}}$$

An epoxide: Ethylene oxide

10.7 Preparation of Epoxides

Oxidation of ethylene by air (oxygen) on a silver catalyst

$$2CH_2=CH_2 + O_2 \xrightarrow[290°C]{260-} 2CH_2-CH_2 \text{ (with O bridging)}$$

ethylene Ag catalyst ethylene oxide

Oxidation of Alkenes with Peroxyacids

$$-\overset{|}{C}=\overset{|}{C}- + C_6H_5CO_2OH \longrightarrow -\underset{|}{C}(O)\underset{|}{C}- + C_6H_5COOH$$

Peroxybenzoic acid Benzoic acid

$$\text{Cyclopentene} + C_6H_5CO_2OH \longrightarrow \text{Cyclopentene oxide} + C_6H_5COOH$$

Cyclopentene Cyclopentene oxide

Displacement of a Halide Ion from a Halohydrin

$$-\underset{X}{\overset{|}{C}}-\underset{OH}{\overset{|}{C}}- + :OH^- \rightarrow -\overset{|}{C}-\overset{|}{C}- \text{ (with O bridging)} + H_2O + :X^-$$

Example

$$CH_3-\underset{OH}{CH}-\underset{Cl}{CH_2} \xrightarrow{\text{conc. Ag. } OH^-} CH_3-CH-CH_2 \text{ (with O bridging)} + H_2O + :Cl^-$$

propylene chlorohydrin

Problem Solving Examples:

Predict the products of the reaction of the following halohydrins with base.

(a) $H-\underset{Br}{\overset{H}{C}}-\underset{OH}{\overset{H}{C}}-CH_3 \longrightarrow$

(b) (cyclohexane ring with OH and Br) $\longrightarrow$

(c) $CH_3-CH_2-\underset{Br}{\overset{H}{C}}-\underset{OH}{\overset{H}{C}}-CH_2CH_3 \longrightarrow$

Halohydrins react with base to yield epoxides.

(a) $H-\underset{Br}{\overset{H}{C}}-\underset{OH}{\overset{H}{C}}-CH_3 + {}^{-}OH \longrightarrow \underbrace{CH_2-CHCH_3}_{O} + H_2O + :Br^-$

(b) (cyclohexane ring with OH and Br) $+ {}^{-}OH \longrightarrow$ (cyclohexane ring fused with O) $+ H_2O + :Br^-$

(c) $CH_3-CH_2-\underset{Br}{\overset{H}{C}}-\underset{OH}{\overset{H}{C}}-CH_2CH_3 + {}^{-}OH \longrightarrow$

$CH_3CH_2-\underbrace{\overset{H}{C}-\overset{H}{C}}_{O}-CH_2CH_3 + H_2O + :Br^-$

Draw the structure of the epoxide(s) produced by epoxidation of (a) 1-hexene and (b) 2-hexene.

One method of synthesis of an epoxide

$$\left(\underbrace{-\overset{|}{C}-\overset{|}{C}-}_{O} \right)$$

is by oxidation of an alkene

$$\left(-\overset{|}{C}=\overset{|}{C}-\right)$$

Peroxyacids are frequently the oxidants used in this epoxidation reaction. It is believed that both carbon-oxygen bonds of the epoxide are formed simultaneously. This means the stereochemistry of the alkene will be retained in the epoxide. Thus, cis alkenes produce cis epoxides and trans alkenes produce trans epoxides. With this in mind, it is possible to draw the structures required.

(a) $CH_3CH_2CH_2CH_2CH{=}CH_2 \longrightarrow CH_3CH_2CH_2CH_2CH{-}CH_2$ (epoxide, with O bridging the last two carbons)

(1-hexene)

(b) 2-hexene exists in cis and trans forms. Hence,

H, H, C_3H_7, CH_3 on C=C $\longrightarrow$ H, H, C_3H_7, CH_3 on C—C bridged by O

(cis-2-hexene)

H, CH_3, C_3H_7, H on C=C $\longrightarrow$ H, CH_3, C_3H_7, H on C—C bridged by O

(trans-2-hexene)

Note how the cis, trans stereochemistry is retained in each case.

10.8 Reactions of Epoxides

Acid-Catalyzed Cleavage

$$-\overset{O}{C}-C- + H^+ \rightleftharpoons -\overset{\overset{H}{O^+}}{C}-C- \xrightarrow{:B^-} -\underset{B}{C}-\overset{OH}{C}-$$

Protonated Epoxide undergoes Nucleophilic Attack

Example

a) Reaction with water yields a glycol.

$$CH_2-CH_2 \text{ (O bridge)} + H_2O \xrightarrow{H^+} \underset{OH}{CH_2}-\underset{OH}{CH_2}$$

Ethylene oxide → Ethylene glycol

b) Reaction with an alcohol yields a compound that is both an ether and an alcohol.

$$CH_2-CH_2 \text{ (O bridge)} + C_6H_5OH \xrightarrow{H^+} C_6H_5-O-CH_2CH_2-OH$$

Ethylene oxide + Phenol → 2-Phenoxyethanol

Base-Catalyzed Cleavage

$$:B^- + -C-C- \text{ (O bridge)} \rightarrow -\overset{B}{C}-\underset{O^-}{C}- \xrightarrow{HB} -\overset{B}{C}-\underset{OH}{C}- + :B^-$$

epoxide undergoes nucleophilic attack

Example

$$C_2H_5O^-Na^+ + CH_2-CH_2 \text{ (O bridge)} \rightarrow C_2H_5OCH_2CH_2OH$$

sodium ethoxide + ethylene oxide → 2-ethoxyethanol

$$NH_3 + CH_2-CH_2 \text{ (O bridge)} \rightarrow H_2NCH_2CH_2OH$$

ethylene oxide → 2-aminoethanol

Reaction with Grignard Reagents

$$R{-}MgX + \underset{O}{CH_2{-}CH_2} \longrightarrow RCH_2CH_2O^-Mg^+X^- \xrightarrow{H^+} RCH_2CH_2OH$$

Primary Alcohol

$$C_6H_5{-}MgBr + \underset{O}{CH_2{-}CH_2} \longrightarrow C_6H_5CH_2CH_2OH$$

Ethylene oxide — 2-Phenylethanol

Problem Solving Examples:

Predict the product(s) of the reaction of ethylene oxide with

(a) H_3O^+

(b) OH^-/H_2O, then H_3O^+

(c) $CH_3S^-Na^+$

(d) $CH_3CH_2CH_2MgBr$, then H_3O^+

This problem focuses on the reactions of epoxides

$$\left(\underset{O}{-\overset{|}{C}-\overset{|}{C}-} \right)$$

which include acid-catalyzed cleavage, base-catalyzed cleavage, and organometallic (e.g., Grignard reagents) cleavage. The general equations for these reactions may be written as follows.

In the presence of acid:

```
                                                                      +
                                                                    Nu-H
           |    |              |    |                          |    |
acid:     -C----C-   H+      -C----C-    :Nu-H               -C --C
            \../    ---->      \../       ------->             |
             O                  +O                            :OH
             ..                  |                             ..
                                 H

                   Nu:
                    |
   - H+       -C --C-
  ------>      |
              :OH
               ..
```

The epoxide is converted by acid (H^+) into the protonated epoxide

```
 (  |    |  )
 ( -C -  C  )
 (   \../   )
 (    O     )
 (   +|     )
 (    H     )
```

which can then undergo attack by any number of nucleophilic reagents in the absence of acid:

```
                                 Nu                    Nu
   |    |                    |    |                |    |
  -C -  C-    :Nu-          -C  -C-      H2O      -C  -C --
    \../     ------->        |          ------>    |    |
     O                     -:O:                   :OH
     ..                      ..                    ..
```

Here, the epoxide, not the protonated epoxide, undergoes nucleophilic attack.

To solve for the products of (a)–(d), substitute into one of the two general equations. The different products reflect the different nucleophilic reagents.

(a) $H_2C{-}CH_2$ (ring with O) + H_3O^+ → $H_2C{-}CH_2$ (ring with $\overset{+}{O}$–H)

Ethylene oxide

$\xrightarrow{H_2O}$ $H_2C(OH){-}CH_2(OH_2^+)$ $\xrightarrow{-H^+}$ $H_2C(OH){-}CH_2(OH)$

(b) $H_2C{-}CH_2$ (ring with O) $\xrightarrow[H_2O]{OH^-}$ $H_2C(O^-){-}CH_2(OH)$ $\xrightarrow{H_3O^+}$ $H_2C(OH){-}CH_2(OH)$

(c) $H_2C{-}CH_2$ (ring with O) $\xrightarrow{CH_3S^-}$ $H_2C(O^-){-}CH_2(SCH_3)$ $\xrightarrow{Na^+}$ $H_2C(O^-Na^+){-}CH_2(SCH_3)$

(d) $H_2C{-}CH_2$ (ring with O) $\xrightarrow{CH_3CH_2CH_2MgBr}$ $H_2C(O^-\overset{+}{Mg}Br){-}CH_2(CH_2CH_2CH_3)$

$\xrightarrow{H_3O^+}$ $H_2C(OH){-}CH_2(CH_2CH_2CH_3)$

Q Write equations for the reaction of ethylene oxide with (a) methanol in the presence of a little H_2SO_4; (b) methanol in the presence of a little $CH_3O^-Na^+$; and (c) aniline.

Ethylene oxide

$$\left(H_2C{-}CH_2 \text{ (ring with O)}\right)$$

is an epoxide, a compound containing the three-membered ring

```
 |     |
-C  -  C-
  \   /
    O
```

Epoxides are highly reactive due to the ease of opening of the highly strained three-membered ring. Epoxides can undergo base and acid-catalyzed cleavage. Reactions in (a)–(c) are examples of such cleavages.

(a) The reaction of ethylene oxide with methanol and H_2SO_4 illustrates the acid-catalyzed cleavage of epoxides. An epoxide when mixed with acid is converted into the protonated epoxide, which can then undergo attack by any of a number of nucleophilic reagents. The general mechanism of the reaction may be written as:

```
     O                          H                OH
    / \                          O+              |
-C  -  C-   +   H+    ⇄      / \             |  |
 |     |                   -C  -  C-        -C--C-
                           ↗|     |          |  |
                        Z:                   Z
```

where Z: represents the nucleophilic reagent. In this problem, methanol is the nucleophilic reagent and attacks the protonated ethylene oxide to produce an alkoxyalcohol (a hydroxyether) as shown:

```
    H
    |
CH3O:   +  H2C - CH2  →  H2C —— CH2   --(-H+)-->   H2C —— CH2
    ..        \ /           |      |                  |      |
               O+         +OCH3   OH                OCH3    OH
               H            H
                                                  2-methoxy-
methanol   protonated                              ethanol
           epoxide
```

Hence, 2-methoxyethanol is the product of the reaction of ethylene oxide with methanol and acid.

(b) Methanol in the presence of a little $CH_3O^-Na^+$ will react with ethylene epoxide to give the same product in (a), 2-methoxyethanol. However, the product comes about by a different mechanism, base-catalyzed cleavage of epoxides. Here, the epoxide itself, not the protonated epoxide, undergoes nucleophilic attack as indicated in the general mechanism on the next page:

$$Z: + -\underset{\diagdown O \diagup}{C - C}- \rightarrow -\overset{Z}{\underset{O^-}{C-C}}- \xrightarrow{HZ} -\overset{Z}{\underset{OH}{C-C}}- + :Z,$$

where Z: denotes the nucleophile. Here, CH_3O^- is the nucleophile that attacks ethylene oxide as shown:

$$CH_3O^- + H_2C \underset{O}{-} CH_2 \rightarrow H_2\overset{OCH_3}{C} — \underset{O^-}{CH_2} \xrightarrow{\text{methanol}}$$

$$H_2\overset{OCH_3}{C} — \underset{OH}{CH_2} + CH_3O^-$$

(2-methoxyethanol)

(c) Aniline,

$$C_6H_5-\ddot{N}H_2$$

is considered to have appreciable basicity due to the pair of nonbonded electrons on the nitrogen atom. Hence, when aniline is added to ethylene epoxide, base-catalyzed cleavage may be expected as shown:

$$C_6H_5-H_2N: + CH_2-CH_2 \text{ (epoxide, O)} \rightarrow C_6H_5-\overset{+}{N}H_2-CH_2CH_2O^- \rightarrow C_6H_5-NH-CH_2CH_2OH$$

Quiz: Aryl Halides, Ethers, and Epoxides

1. What are the missing reactants in the equation?

$$\text{4-ClC}_6\text{H}_4\text{SO}_3\text{H} \xrightarrow{?} \text{4-Cl-3,5-(NO}_2)_2\text{C}_6\text{H}_2\text{SO}_3\text{H} \xrightarrow{?} \text{4-NH}_2\text{-3,5-(NO}_2)_2\text{C}_6\text{H}_2\text{SO}_3\text{H}$$

(A) NH_3 followed by H_2SO_4

(B) H_2SO_4/HNO_3 followed by NH_3/NH^-_2

(C) $NaNO_2/HCl$ followed by NH_3/NH^-_2

(D) $(CH_3CO)_2O$ followed by NH_3/NH^-_2

(E) None of the above.

2. Predict the product resulting from the reaction between ethylene oxide and the gringnard reagent Ethyl Magnesium Bromide.

$$H_2C{-}CH_2\ (\text{epoxide, bridged by O}) \quad + \quad CH_3CH_2MgBr$$

(A) $Br\,CH_2\,CH_2\,Br$

(B) $CH_3\,CH_2O\,CH_2\,CH_2\,O\,CH_2\,CH_3$

(C) $CH_3\,CH_2\,O\,CH_2\,CH_2\,Br$

(D) $Br\,CH_2\,CH_2\,OH$

(E) $CH_3\,CH_2\,\text{-}O\,CH_2\,CH_2\,OH$

3. Which of the following aryl halides is named m-bromo-toluene?

(A)

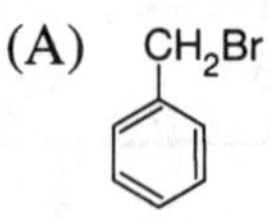

(D)

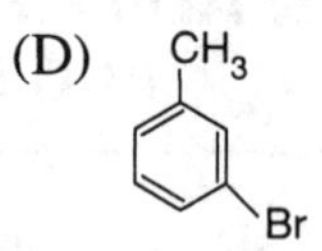

(B) CH_3 Br

(E) Br

(C) CH_3 Br

4. Physical properties of aryl halides include all of the following EXCEPT

   (A) soluble in organic solvents but not soluble in water.

   (B) colorless liquids.

   (C) para isomers have a melting point higher than ortho isomers.

   (D) isomeric dihalobenzenes have similar boiling points.

   (E) purification is by recrystallization.

5. Aryl halides may be prepared by all of the following EXCEPT

   (A) Sandmeyer reaction.

   (B) substitution.

   (C) addition.

   (D) Friedel-Crafts alkylation.

   (E) nitrogen replacement of a diazonium salt.

6. The method of preparing an ether as shown is known as

$$CH_2=CHCH_2Cl + NaOCH_2CH_3 \longrightarrow CH_2{=}CHCH_2OCH_2CH_3 + NaCl$$

(A) Williamson reaction.

(B) dehydration.

(C) alkoxymercuration-demercuration.

(D) sodium substitution.

(E) Wurtz reaction.

7. All of the following may react with ethers EXCEPT

(A) H_2SO_4.

(B) PCl_5.

(C) H_2O.

(D) HI.

(E) Al_2O_3.

8. What is the product of the following reaction?

$$CH_3-CH_2-O-CH_3 + Br_2 \rightarrow$$

(A) $BrCH_2-CH_2-O-CH_3$

(B) $CH_3-CH_2-O-CH_2Br$

(C) $CH_3-CHBr-O-CH_3$

(D) $CH_3-CH_2-O-BrCH_2$

(E) $BrCH_2-CH_2-O-CH_2Br$

9. All of the following may react with epoxides EXCEPT

(A) phenol.

(B) Grignard reagents.

(C) water.

(D) ethoxides.

(E) carbon tetrachloride.

10. What is the product of the reaction of cyclohexene with peroxybenzoic acid?

(A) 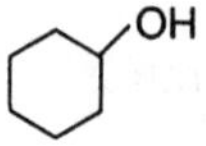 OH

(B) 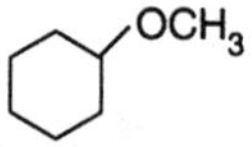 OCH_3

(C) 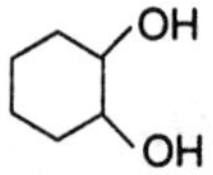 OH OH

(D) 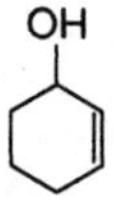 OH

(E) 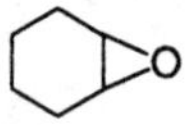 O

ANSWER KEY

1.	(B)	6.	(A)
2.	(E)	7.	(E)
3.	(D)	8.	(C)
4.	(A)	9.	(E)
5.	(D)	10.	(E)

CHAPTER 11

Alcohols and Glycols

Alcohols are hydrocarbon derivatives in which one or more hydrogen atoms have been replaced by the –OH (hydroxyl) group. They have the general formula R – OH, where R may be either alkyl or aryl.

11.1 Nomenclature (IUPAC System)

Alcohols are named by replacing the "-e" ending of the corresponding alkane with the suffix "-ol." The alcohol may also be named by adding the name of the R group to the same alcohol.

Example CH_3CH_2OH Ethanol or ethyl alcohol

Depending on what carbon atom the hydroxyl group is attached to, the alcohol is prefixed as follows:

A) Primary (–OH attached to 1° carbon) alcohols are prefixed "n-" or "1-."

B) Secondary (–OH attached to 2° carbon) alcohols are prefixed "sec-" or "2-."

C) Tertiary (–OH attached to 3° carbon) alcohols are prefixed "tert-" or "3-."

Example

$CH_3CH_2CH_2CH_2OH$

n- or 1- butanol

$$\begin{array}{c} CH_3CH_2CHCH_3 \\ | \\ OH \end{array}$$

sec- or 2-butanol

$$\begin{array}{c} CH_3 \\ | \\ CH_3CH_2-C-CH_3 \\ | \\ OH \end{array}$$

tert- or 3- pentanol

Problem Solving Examples:

What are alcohols and how are they classified?

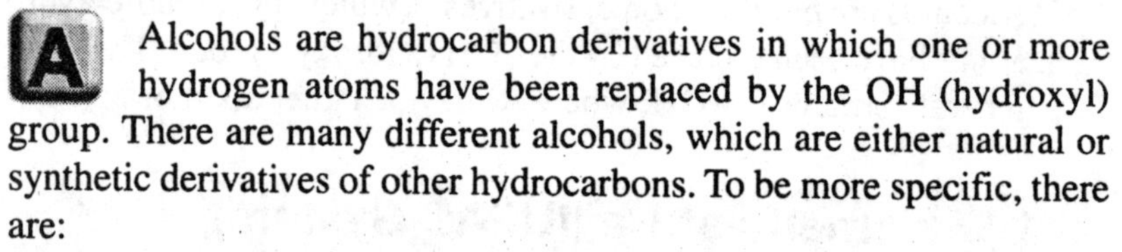

Alcohols are hydrocarbon derivatives in which one or more hydrogen atoms have been replaced by the OH (hydroxyl) group. There are many different alcohols, which are either natural or synthetic derivatives of other hydrocarbons. To be more specific, there are:

I. Monohydroxy Alcohols

  A. Primary
  B. Secondary
  C. Tertiary

II. Dihydroxy Alcohols (glycols)

III. Trihydroxy Alcohols

IV. Polyhydroxy Alcohols

The monohydroxy alcohols are represented by the general formula $R-OH$ or $C_nH_{2n+1}OH$.

Specific examples of each are:

Primary: $CH_3CH_2CH_2CH_2OH$ 1-butanol (butyl alcohol)

Secondary: $\begin{array}{c} CH_3CH_2CH-CH_3 \\ | \\ OH \end{array}$ 2-butanol (sec-butyl alcohol)

```
                CH3
                |
Tertiary:  CH3-C -CH3
                |
                OH
```

2-metyhyl-2-propanol
(tert-butyl alcohol)

Dihydroxy and trihydroxy alcohols have two and three hydroxyl groups, respectively, on their hydrocarbon molecule. For example:

```
(a)   CH2-CH2
      |    |
      OH   OH
```

1,2-ethanediol (ethylene glycol) “dihydroxy alcohol”

```
(b)   CH2-OH
      |
      CH-OH
      |
      CH2-OH
```

1,2,3-propanetriol (glycerol) “trihydroxy alcohol”

These alcohols that contain more than one hydroxyl group may be represented by the general formula C_nH_{2n+2-y} $(OH)_y$, where y is two or more. The common polyhydroxy alcohols usually have one “OH” group on every carbon atom in the molecule.

Name the following alcohols using the IUPAC system of nomenclature:

(a) $(CH_3)_2CHCH_2OH$

```
                        OH
                        |
(b) CH2  =  CHCH2CH-CH3
```

```
                              Cl
                              |
(c) CH3CH  -CH-CH-CHCH2OH
        |       |          |
        CH2   CH3        CH3
        |
        CH3
```

(d) 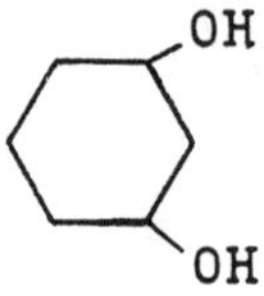 [cyclohexane ring with OH at positions 1 and 3]

OH

OH

(e) 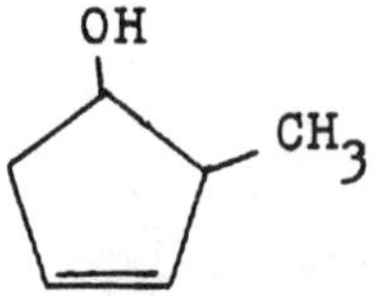 [cyclopentene ring with OH and CH_3]

OH

CH_3

A An alcohol is named as a derivative of the longest continuous carbon chain containing the ^-OH group. The chain is numbered so as to give the hydroxyl group the lowest number, and the positions of substituents are indicated by number. The -ol ending is added to designate an ^-OH group. For cyclic compounds, the same rules hold true; thus, the first carbon number in the ring is the one that possesses the hydroxyl functional group. With these rules in mind, the alcohol compounds can be named.

(a) $(CH_3)_2CHCH_2OH$ 2-methyl-1-propanol

(b) $CH_2 = CHCH_2CH(OH)-CH_3$ 4-penten-2-ol

(c) $CH_3CH(CH_2CH_3)-CH(CH_3)-CH(Cl)-CH(CH_3)CH_2OH$ 3-chloro-2,4,5-trimethyl-1-heptanol

(d) 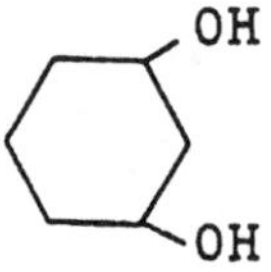 [cyclohexane ring with OH at positions 1 and 3] 1,3-cyclohexanediol

OH

OH

(e) OH, CH_3 2-methyl-3-cyclopenten-1-ol

11.2 Physical Properties of Alcohols

A) Alcohols are high boiling.

B) Lower alcohols are soluble in water.

C) Refractive index, density, and boiling point of alcohols increase with an increase in C-content, but solubility in water decreases.

D) The introduction of additional –OH groups into a molecule tends to intensify the characteristic properties of the hydroxyl group.

E) In general, for a given number of carbon atoms, the boiling point, the sweetness, and the solubility of an alcohol in water increase with the number of hydroxyl groups.

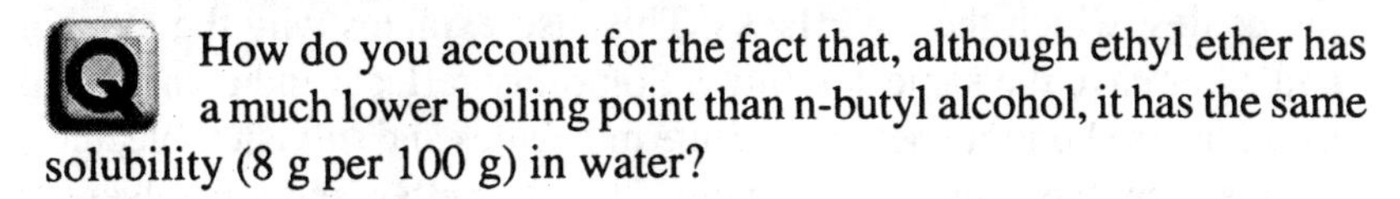

Q How do you account for the fact that, although ethyl ether has a much lower boiling point than n-butyl alcohol, it has the same solubility (8 g per 100 g) in water?

A Examination of the structures of ethyl ether and n-butyl alcohol shows that they are isomers; they are different compounds with the same molecular formula ($C_4H_{10}O$).

$CH_3CH_2CH_2CH_2OH$	$CH_3CH_2OCH_2CH_3$
n-butyl alcohol	ethyl ether

A consideration of what factors influence boiling points and solubility will aid in the solution of this problem. Associated liquids are ones whose molecules are held together by hydrogen bonds. It takes considerable energy to break these hydrogen bonds. Hence, the boiling point is raised by hydrogen bonding between like molecules of a compound. It can be seen from the structure of n-butyl alcohol that the alcohol molecules can form hydrogen bonds to each other and to water molecules. The ether can form hydrogen bonds to only the water, however. The situation is depicted below.

```
                         H
                        /
CH3CH2CH2CH2OH-H---O          ⇐ H-bonding with water
                        \
                         H

                        H
                       /
CH3CH2CH2CH2O-H---OCH2CH2CH2CH3  ⇐ H bonding to each
                                              other

CH3CH2OCH2CH3           ⇐ H-bonding with water only.
       :
       :
       H
        \
         O
         |
         H
```

This means the ether must have the lower boiling point. The boiling point is determined, in part, by the hydrogen bonding between (like) molecules, which the ether lacks. This also explains why the alcohol and ether have the same solubility. Solubility reflects and is increased by hydrogen bonding between solute molecules and solvent molecules. In this case, the solute is either the alcohol or ether and the solvent is water. As shown above, both the alcohol and ether may participate in hydrogen bonding with water equally well. Hence, they should possess the same solubility.

11.3 Preparation of Monohydroxy Alcohols

A) Substitution of –OH for –X when alkyl halides are treated with aqueous sodium or potassium hydroxide.

$$R - X + NaOH(aq) \rightarrow R\text{–}OH + NaX$$

Example

$CH_3CH_2 - Cl + NaOH(aq) \rightarrow$

chloroethane $\qquad$ $CH_3CH_2 - OH + NaCl$

ethanol

B) Introduction of alkyl and hydrogen groups into a carbonyl or epoxy compound upon treatment with a Grignard reagent in dry ether and subsequent hydrolysis.

a) $H - CHO + R - Mg - X \rightarrow R - CH_2 - O - Mg - X$

$R - CH_2 - O - Mg - X + HX(aq) \rightarrow R - CH_2 - OH + MgX_2$

primary alcohol

Example

$$H_2C{=}O + CH_3{-}Mg{-}I \rightarrow CH_3{-}CH_2{-}O{-}Mg{-}I$$

formaldehyde

$CH_3 - CH_2 - O - Mg - I + HCl(aq) \rightarrow CH_3 - CH_2 - OH + Mg - I - Cl$

b) 1. $R - CHO + R' - MgX + HX(aq) \rightarrow R - CHOH - R' + MgX_2$

secondary alcohol

2. $RCH{-}CH_2$ (epoxide, O bridging) $+ R'{-}Mg{-}X + HX(aq) \rightarrow R{-}CHOH{-}CH_2{-}R'$

$+ MgX_2$

c) 1. $R_2C = O + R'MgX + 2HX(aq) \rightarrow R_2R'C - OH + MgX_2$

tertiary alcohol

2. $R_2C{-}CH_2$ (epoxide, O bridging) $+ R_2'Mg + HX(aq) \rightarrow R_2C(OH){-}CH_2{-}R'$

$+ MgX_2 + R'H$

C) By the catalytic hydrogenation or by the reduction of aldehydes or ketones in acid solution.

a) $R - CHO + Zn + 2H_2O$, acidic $\rightarrow R - CH_2 - OH + Zn^{++} + 2OH^-$

aldehyde $\qquad$ 1° alcohol

Example

$$CH_3-CH_2-CHO + Zn + 2H_2O, \text{acidic} \rightarrow$$
$$CH_3CH_2-CH_2-OH + Zn^{++} + 2OH^-$$

1–propanol

b) $R_2C = O + Zn + 2H_2O, \text{acidic} \rightarrow R_2CHOH + Zn^{++} + 2OH^-$

ketone 2° alcohol

D) Hydration of alkenes

$$R'-CH_2-CH_2-CH_2-R + \text{cracking} \rightarrow R'-CH=CH_2 + CH_3-R$$

R > R'

$$R'-CH=CH_2 + H_2SO_4 \rightarrow R'-CH(OSO_2-OH)-CH_3$$

$$\xrightarrow[\text{excess}]{H_2O} R'-CH(OH)-CH_3 + H_2SO_4$$

E) By the action of heat and pressure on a mixture of carbon monoxide or carbon dioxide and hydrogen in the presence of a catalyst (zinc chromite).

$$CO + 2H_2 \xrightarrow[400-500°C, 200 \text{ atm.}]{\text{zinc chromite}} CH_3OH$$

methanol

F) Reaction of amines with nitrous acid.

a) $CH_3-NH_2 + HO-NO \xrightarrow[H^+]{NaNO_2} CH_3OH + N_2 + H_2O$

Methyl Amine Methanol

b) $CH_3\text{-}CH_2\text{-}NH_2 + HO\text{-}NO \xrightarrow[H^+]{Na\text{-}NO_3} CH_3\text{-}CH_2\text{-}OH + N_2 + H_2O$

Ethyl Amine → Ethanol

G) Oxymercuration-demercuration

$$>C{=}C< + H_2O \xrightarrow{Hg(OAc)_2} -\underset{OH}{\underset{|}{\overset{|}{C}}}-\underset{HgOAc}{\underset{|}{\overset{|}{C}}}- \xrightarrow{NaBH_4} -\underset{OH}{\underset{|}{\overset{|}{C}}}-\underset{H}{\underset{|}{\overset{|}{C}}}-$$

Markovnikov addition

Example

$$CH_3-\underset{CH_3}{\underset{|}{\overset{CH_3}{\overset{|}{C}}}}-CH{=}CH_2 \xrightarrow[H_2O]{Hg(OAc)_2} \xrightarrow{NaBH_4} CH_3-\underset{CH_3}{\underset{|}{\overset{CH_3}{\overset{|}{C}}}}-\underset{OH}{\underset{|}{CH}}-CH_3$$

3,3-dimethyl-1-butene → 3,3-dimethyl-2-butanol

H) Hydroboration-oxidation

$$>C{=}C< + (BH_3)_2 \rightarrow -\underset{H}{\underset{|}{\overset{|}{C}}}-\underset{B<}{\underset{|}{\overset{|}{C}}}- \xrightarrow[OH^-]{H_2O_2} -\underset{H}{\underset{|}{\overset{|}{C}}}-\underset{OH}{\underset{|}{\overset{|}{C}}}- + B(OH)_3$$

diborane; alkylborane; anti-Markovnikov addition

Example

$$CH_3\underset{CH_3}{\underset{|}{\overset{CH_3}{\overset{|}{C}}}}-CH{=}CH_2 \xrightarrow{(BH_3)_2} \xrightarrow{H_2O_2OH^-} CH_3-\underset{CH_3}{\underset{|}{\overset{CH_3}{\overset{|}{C}}}}-CH_2-CH_2OH$$

3,3-dimethyl-1-butene → 3,3-dimethyl-1-butanol

Problem Solving Examples:

Q What are the various methods of preparing alcohols?

A There are many ways of preparing alcohols through organic reactions. As yet, it is not possible to introduce hydroxyl groups directly into alkanes, and all synthetic reactions of alcohols start from compounds containing reactive functional groups. The most common methods of preparing alcohols are as follows:

(1) Hydration of olefins in the presence of acid.

(2) Hydrolysis of alkyl halides by water or alkali.

(3) Hydrolysis of ethers in strongly acidic conditions.

(4) Hydrolysis of esters

  (a) acid catalyzed hydrolysis

  (b) alkaline hydrolysis (saponification)

(5) Reduction or catalytic hydrogenation of aldehydes, ketones, carboxylic acids, and esters.

Show the product of each compound by hydrolysis of alkyl halides:

a) $CH_3CH_2CH_2CH_2CH_2Br$ (b) C_6H_5–CH_2Cl

n-bromopentane Benzyl chloride

A This method of preparing an alcohol can be done by adding a concentrated aqueous alkali to the reactant and reflux the mixture. The alkali usually used is sodium hydroxide (NaOH) or potassium hydroxide (KOH) in which the –OH group replaces the halogen on the molecule. The general reaction for this is:

$$RX + KOH\ (aqueous) \rightarrow R-OH\ +\ K^+X^-.$$

Alkyl halide Alkyl alcohol

The R shows the alkyl functional group, and the X signifies the halogen (Br or Cl). With the above facts in mind, the products of the two precursors can be derived by hydrolysis.

(a)

$$CH_3CH_2CH_2CH_2CH_2Br \xrightarrow{NaOH\ (aqueous)} CH_3CH_2CH_2CH_2CH_2OH + Na^+Br^-$$

n-Bromopentane n-Pentanol

(b)

$$C_6H_5CH_2Cl + KOH\ (aqueous) \rightarrow C_6H_5CH_2OH + K^{+}Cl^{-}$$

Benzyl chloride → Benzyl alcohol

11.4 Reaction of Monohydroxy Alcohols

A) Replacement of the hydrogen atom of the hydroxyl group when treated with:

a) Active metals

$$2R-OH + 2Na \rightarrow 2R-)O^{-}\overset{+}{Na} + H_2$$

b) Acid halides

$$R-O\boxed{H + X}OC-R' \rightarrow R-O\overset{O}{\overset{\|}{C}}-R' + HX$$

c) Organic acids

$$R-O\boxed{H + HO}OC-R' \xrightarrow[\text{or } P_2O_5]{H_2SO_4} R-O\overset{O}{\overset{\|}{C}}-R' + H_2O$$

d) Alkyl hydrogen sulfates

$$R-O\boxed{H + HOSO_2O}R' \rightarrow R-OR' + H_2SO_4$$

e) Grignard reagent

$$R-O\boxed{H + R}-Mg-X \rightarrow R-O-Mg-X + RH$$

B) Replacement of the hydroxyl group when treated with:

a) Hydriodic acid/(red phosphorus)

$$R-\boxed{OH + H}I \rightarrow RI + H_2O$$

b) Hydrobromic acid, dry (or hydrochloric acid)

$$R-\boxed{OH + H}Br \rightarrow R-Br + H_2O$$

c) Hydrochloric acid (or HBr)/conc. sulfuric acid

$$R-\boxed{OH + H}Cl \rightarrow R-Cl + H_2O$$

d) Sulfuric acid

$$R-\boxed{OH + H}OSO_2-OH \rightarrow H_2O + R-O-SO_2-OH$$

e) Nitric acid

$$R-\boxed{OH + H}NO_3 \rightarrow H_2O + R-NO_3$$

f) Phosphorus trihalides

$$3R - OH + PX_3 \rightarrow P(OH)_3 + 3RX$$

g) Phosphorus pentahalides

$$R - OH + PCl_5 \rightarrow R - CL + HCL + POCl_3$$

C) Dehydration of alcohols by acids to give unsaturated derivatives.

$$R - CHOH - CH_2 - R' + P_2O_5 \rightarrow R - HC = CH - R' + 2HPO_3$$

Alkene

D) Oxidation of alcohols, to give derivatives, when treated with:

a) One mole of dichromic acid per three moles of $R - CH_2 - OH$.

$$R - CH_2 - OH + \text{oxidation} \rightarrow R - CHO \rightarrow R - CO - OH \rightarrow \rightarrow \text{cleavage}$$

aldehyde — oxidized derivatives

Example

$$3CH_3 - CH_2 - OH + Na_2Cr_2O_7/4H_2SO_4$$
$$\rightarrow 3CH_3 - CHO + Na_2SO_4 + Cr_2(SO_4)_3 + 7H_2O$$

acetaldehyde

b) Two moles of dichromic acid per three moles of $R - CH_2 - OH$.

Example

$$3CH_3 - CH_2 - OH + 2Na_2Cr_2O_7/8H_2SO_4$$
$$\rightarrow 3CH_3 - CO - OH + 2Na_2SO_4 + 2Cr_2(SO_4)_3 + 11H_2O$$

Acetic acid

c) One mole of dichromic acid per three moles of $R_2 - CH - OH$.

$$R_2 - CH - OH + \text{oxidation} \rightarrow R_2C = O \rightarrow \text{oxidized derivative} \rightarrow \text{cleavage}$$

ketone

Primary alcohols can be oxidized to form aldehydes and acids.

Example $CH_3CH_2CH_2OH + KMnO_4 \rightarrow CH_3CH_2COOH$

n-propanol propionic acid

$$CH_3CH_2CH_2OH + K_2Cr_2O_7 \rightarrow CH_3CH_2\overset{\overset{\large H}{|}}{C}=O$$

propionaldehyde

Example

Secondary alcohols can be oxidized to form ketones.

$$CH_3CH_2\overset{\overset{\large CH_3}{|}}{C}HOH \xrightarrow[\text{or } CrO_3]{K_2Cr_2O_7} CH_3-CH_2\overset{\overset{\large CH_3}{|}}{C}=O$$

isobutyl alcohol
2-hydroxy butane

methyl ethyl ketone

Tertiary alcohols cannot be oxidized.

Problem Solving Examples:

Predict the products of each alcohol if a hydrogen halide were added.

(a)

$$CH_3-\overset{\overset{\large CH_3}{|}}{\underset{\underset{\large OH}{|}}{C}}-CH_3$$

t-Butyl alcohol

(b) Cyclohexanol (cyclohexane ring bearing OH)

(c)

$$CH_3-\overset{\overset{\large CH_3}{|}}{\underset{\underset{\large CH_3}{|}}{C}}-CH_2OH$$

Neopentyl alcohol

(d) $CH_3CH_2CH_2CH_2CH_2OH$

n-Pentyl alcohol

The chemical properties of an alcohol, ROH, are determined by its functional group, –OH, the hydroxyl group.

Reactions of an alcohol can involve the breaking of either of two bonds: the C – OH bond, with removal of the –OH group; or the O – H bond, with removal of –H. Either kind of reaction can involve substitution, in which a group replaces the –OH or –H, or elimination, in which a double bond is formed.

Alcohols react readily with hydrogen halides to yield alkyl halides and water. The reaction is carried out either by passing the dry hydrogen halide gas into the alcohol, or by heating the alcohol with the concentrated aqueous acid. Sometimes hydrogen bromide is generated in the presence of the alcohol by reaction between sulfuric acid (H_2SO_4) and sodium bromide.

The general formula for this type of reaction is:

$$R - OH + HX \rightarrow RX + H_2O$$

and by using this formula with the above principles, the products of the given reactions may be predicted.

(a) $(CH_3)_3C{-}OH \xrightarrow[\text{room temp.}]{\text{conc. HCl}} (CH_3)_3C{-}Cl + CH_2{=}C(CH_3)_2$

t-Butyl alcohol → t-Butyl chloride + Isobutylene

(b) Cyclohexanol (cyclohexane ring with –OH) $\xrightarrow{\text{Dry HBr}}$ Cyclohexyl bromide (cyclohexane ring with –Br) + Cyclohexene

Cyclohexanol → Cyclohexyl bromide + Cyclohexene

(c) $CH_3{-}C(CH_3)_2{-}CH_2OH + HCl \rightarrow CH_3{-}CCl(CH_3){-}CH_2{-}CH_3$

Neopentyl alcohol → t-Pentyl chloride

From solution (c) it is seen that the halogen does not always become attached to the carbon that originally held the hydroxyl group. Even the carbon skeleton may be different from that of the starting material. The reason is because compounds during reaction seek their most stable form, and they will rearrange to that form if it is energetically feasible.

Neopentyl alcohol arranged as shown:

$$CH_3-\overset{\displaystyle CH_3}{\underset{\displaystyle CH_3}{\overset{|}{\underset{|}{C}}}}-CH_2OH \xrightarrow{H^+} CH_3-\overset{\displaystyle CH_3}{\underset{\displaystyle CH_3}{\overset{|}{\underset{|}{C}}}}-CH_2\overset{+}{O}H_2 \xrightarrow{-H_2O}$$

$$CH_3-\overset{\displaystyle CH_3}{\underset{\displaystyle CH_3}{\overset{|}{\underset{|}{C}}}}-CH_2^+ \rightarrow CH_3-\overset{+}{\underset{\displaystyle CH_3}{\underset{|}{C}}}-CH_2CH_3 \xrightarrow{Cl^-} CH_3\overset{\displaystyle Cl}{\underset{\displaystyle CH_3}{\overset{|}{\underset{|}{C}}}}CH_2CH_3$$

(d)

$$\underset{\text{n-Pentyl alcohol}}{CH_3CH_2CH_2CH_2CH_2OH} + HCl \xrightarrow[\text{heat}]{ZnCl_2} \underset{\text{n-Pentyl chloride}}{CH_3CH_2CH_2CH_2CH_2Cl} + H_2O$$

Q Sodium metal was added to tertbutyl alcohol and allowed to react. When the metal was consumed, ethyl bromide was added to the resulting mixture. Work-up of the reaction mixture yielded a compound of formula $C_6H_{14}O$.

In a similar experiment, sodium metal was allowed to react with ethanol. When tertbutyl bromide was added, a gas was evolved, and work-up of the remaining mixture gave ethanol as the only organic material.

(a) Write equations for all reactions. (b) What familiar reaction type is involved in each case? (c) Why did the reactions take different courses?

A (a), (b) It is known that alcohols may act as acids. Alcohols may react with active metals to form the alkoxide ion, RO^-, and hydrogen gas. The reaction may be written as:

$$RO-H+M \rightarrow RO^-M^+ + {}^1\!/\!_2H_2,$$

where M = Na, K, Mg, Al, etc., and the reactivity of ROH is $CH_3OH > 1° > 2° > 3°$. Hence, when sodium metal is added to tertbutyl alcohol and allowed to react, sodium tert-butoxide is formed as shown:

$$CH_3\text{-}\underset{\substack{|\\ OH}}{\overset{\substack{CH_3\\ |}}{C}}\text{—}CH_3 + Na \longrightarrow CH_3\text{-}\underset{\substack{|\\ O^-Na^+}}{\overset{\substack{CH_3\\ |}}{C}}\text{—}CH_3 + \tfrac{1}{2}H_2$$

t-butyl alcohol   sodium t-butoxide

Alkoxides are useful reagents; they are used as powerful bases (stronger than hydroxide). Sodium t-butoxide and ethyl bromide (CH_3CH_2Br) are mixed together to give $C_6H_{14}O$. This reaction illustrates nucleophilic substitution. Ethyl bromide is an alkyl halide, and nucleophilic substitution is typical of such organic compounds. Nucleophilic substitution may be generalized as

$$R\text{–}X + :Z \rightarrow R\text{-}Z + :X^-$$

A nucleophilic reagent   Leaving group

As mentioned, sodium tert-butoxide is a strong base. It is an electron-rich species, so that it may be described as a nucleophile. Consequently, the equation for the reaction with ethyl bromide becomes:

$$CH_3\text{-}\underset{\substack{|\\ CH_3}}{\overset{\substack{CH_3\\ |}}{C}}\text{—}O^-Na^+ + CH_3CH_2Br \rightarrow CH_3\text{-}\underset{\substack{|\\ CH_3}}{\overset{\substack{CH_3\\ |}}{C}}\text{—}OCH_2CH_3 \quad (C_6H_{14}O)$$

t-butyl ethyl ether

In a similar experiment, another alkoxide, sodium ethoxide, is obtained when sodium is allowed to react with ethanol. The reaction may be written as follows:

$$CH_3CH_2OH + Na \rightarrow CH_3CH_2O^-Na^+ + \tfrac{1}{2}H_2$$

Due to the fact that the only organic compound present after tertbutyl bromide is added to sodium ethoxide is ethanol, this reaction cannot be the same as the one discussed above (when ethanol was added to sodium t-butoxide). In fact, this reaction is one of elimination. Recall that in the dehydrohalogenation to produce alkenes, the base (OH^-) pulled a hydrogen ion away from carbon, and simultaneously a halide ion separated. This is exactly what happens when sodium ethoxide and t-butyl bromide are mixed together, except that in this case the sodium ethoxide replaces OH^- as the base. The reaction may be written as:

$$CH_3CH_2O^-Na^+ + CH_3-\underset{\underset{Br}{|}}{\overset{\overset{CH_3}{|}}{C}}-CH_3 \rightarrow \underset{\text{(isobutylene)}}{CH_3-\overset{\overset{CH_3}{|}}{C}=CH_2} + \underset{\text{Ethanol}}{CH_3CH_2OH} + NaBr$$

Isobutylene, the alkene produced, is a gas and diffuses away. This leaves only NaBr (inorganic) and ethanol.

(c) How can the different courses of the two reactions be explained? Examine the nature of the alkyl halide involved in each case. In the first case, ethyl bromide, a primary alkyl halide, was used. In the second case, tertbutyl bromide, a tertiary alkyl halide, was used. Both nucleophilic substitution and elimination are typical reactions of alkyl halides—they can compete with each other. But note the reactivities for each:

E_2 or E_1 elimination: $3° > 2° > 1°$

S_N2 (Substitution nucleophilic bimolecular):

$CH_3X > 1° > 2° > 3°$

Hence, when the primary alkyl halide is used, S_N2 is favored over elimination. Therefore, nucleophilic (S_N2) substitution occurs when ethyl bromide is added to sodium t-butoxide. However, when the tertiary alkyl halide is employed, the elimination reaction predominates, as seen when t-butyl bromide is added to sodium ethoxide.

11.5 Uses of Alcohols

A) Alcohols are widely used in synthesis, especially in that of ester, and as solvents.

B) Methanol is used as an antifreeze and in the production of methanal (formaldehyde), which is used in the synthesis of resins.

C) Ethanol is used as a solvent, synthetic intermediate, antifreeze, and as an ingredient in alcoholic beverages.

D) Butyl and amyl alcohols are used in the preparation of esters for the lacquer industry.

11.6 Glycols

Alcohols containing more than one hydroxyl group (polyhydroxyalcohols) are represented by the general formula $CnH_{2n+2-y}(OH)_y$. Polyhydroxyalcohols containing two hydroxyl groups are called glycols or diols.

Example 1,3-butanediol

$$CH_3-\underset{\displaystyle OH}{\underset{|}{CH}}-CH_2-CH_2-OH$$

11.7 Preparation of Glycols

Treatment of ethylene with hypochlorous acid and subsequent hydrolysis.

$$H_2C=CH_2 + HO-Cl, \text{aq.} \rightarrow \underset{\text{halohydrin}}{HO-CH_2-CH_2-Cl} + NaHCO_3, \text{aq}$$

$$\rightarrow HO-CH_2-CH_2-OH \text{ (glycol)} + NaCl + CO_2$$

Oxidation of ethylene with the presence of gold or silver and the addition of water.

$$H_2C=CH_2 + \tfrac{1}{2}O_2 \xrightarrow{Au/Ag} \underset{\text{epoxide}}{H_2C\overset{O}{\text{—}}CH_2}$$

$$\xrightarrow{+H_2O} HO-CH_2-CH_2-OH$$

Cold dilute potassium permanganate causes hydroxylation of alkenes to yield glycols.

$$>C=C< + \text{cold, dilute } KMnO_4 \rightarrow >C(OH)-C(OH)<$$

$$\text{cyclohexene} + \text{dil. } KMnO_4 \xrightarrow{5^\circ C} \text{cis-1,2-cyclohexanediol}$$

cis-1,2-cyclo-
hexanediol

Aldol condensation of α,α-dialkylacetaldehydes with formaldehyde.

$$R_2CHCHO + CH_2O \xrightarrow{KOH} HOCH_2C(R)_2CH_2OH$$

Reduction of dicarbonyl compounds.

Example

$$CH_3C(=O)(CH_2)_3C(=O)CH_3 + NaBH_4 \xrightarrow{C_2H_5OH} CH_3CH(OH)(CH_2)_3CH(OH)CH_3$$

2,6-heptanediol

Cleavage of epoxides by water in the presence of mineral acids to give trans-glycol.

$$>C=C< + >C\overset{:O:}{-}C< \underset{-H_2O}{\overset{+H_3O}{\rightleftharpoons}} >C\overset{\ddot{O}^+H}{-}C< \xrightarrow{+H_2O} H_2\ddot{O}:\rightarrow C-C(\ddot{O}H)$$

$$\rightarrow (HO)C-C(\overset{+}{O}H_2) \xrightarrow{-H^+} -C(OH)-C(OH)-$$

Hydrolysis of alkyl dihalides.

$$\underset{\substack{| \\ X}}{R-CH}-\underset{\substack{| \\ X}}{CH}-R + 2{:}OH^- \xrightarrow[H_2O]{Na_2CO_3} \underset{\substack{| \\ OH}}{R-CH}-\underset{\substack{| \\ OH}}{CH}-R + 2{:}X^-$$

where X = Cl,Br

Hydrolysis of halohydrins in the presence of bases.

$$\underset{\substack{| \\ X}}{R-CH}-\underset{\substack{| \\ OH}}{CH}-R + {:}OH^- \rightarrow \underset{\substack{| \\ OH}}{R-CH}-\underset{\substack{| \\ OH}}{CH}-R + {:}X^-$$

Problem Solving Example:

Q Propylene oxide can be converted into propylene glycol by the action of either dilute acid or dilute base. When optically active propylene oxide is used, the glycol obtained from acidic hydrolysis has a rotation opposite to that obtained from alkaline hydrolysis. What is the most likely interpretation of these facts?

Propylene oxide,

$$H_2C\overset{O}{\frown}C\begin{smallmatrix} CH_3 \\ H \end{smallmatrix}$$

is an epoxide because it contains the three-membered ring containing oxygen

$$-\overset{|}{C}\underset{O}{\smile}\overset{|}{C}-$$

Due to the ease of opening this strained ring, it is highly reactive. Dilute acid or base can hydrolyze it into propylene glycol;

$$CH_3\underset{\substack{| \\ OH}}{CH}\underset{\substack{| \\ OH}}{CH_2}$$

If optically active propylene glycol is used, depending on the method of hydrolysis, two different enantiomers are formed. Enantiomers are mirror-image isomers whose chemical and physical properties are the same, except for the ability to rotate plane polarized light in opposite directions.

Acid hydrolysis begins with the protonation of the epoxide (see the diagram). A water molecule then attacks a carbon adjacent to the oxygen. It attacks the more highly substituted carbon. In this case, the carbon attacked is a chiral carbon, that is, a carbon with four different atoms or groups of atoms. Since the water attacks at the chiral carbon, inversion of configuration results to change the rotation of plane polarized light.

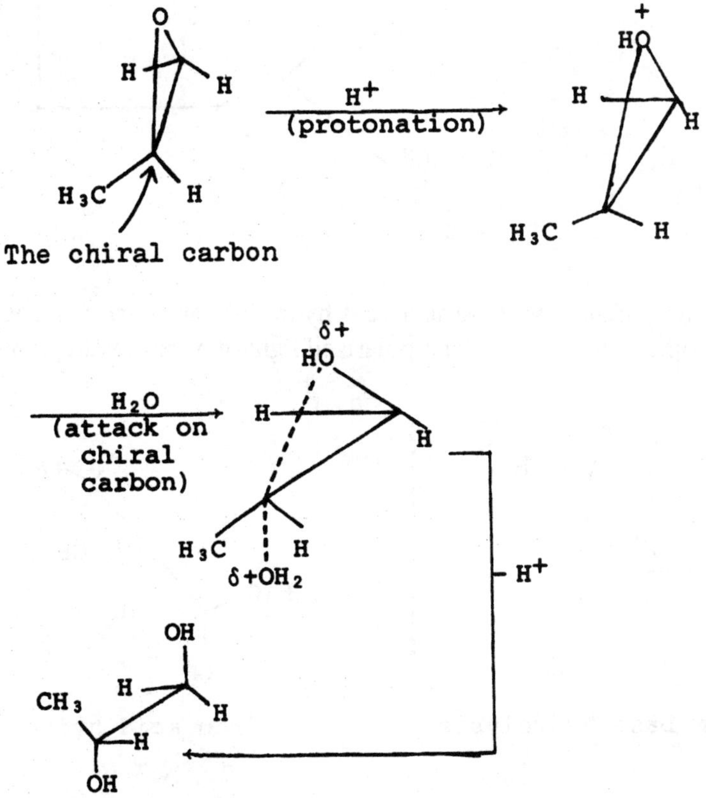

The chiral carbon

H^+
(protonation)

H_2O
(attack on chiral carbon)

H^+

Base hydrolysis takes place with retention of configuration. The attack of hydroxide is at the 1° carbon, where no bond to the chiral carbon is broken. Hence, the rotation of plane polarized should not be altered.

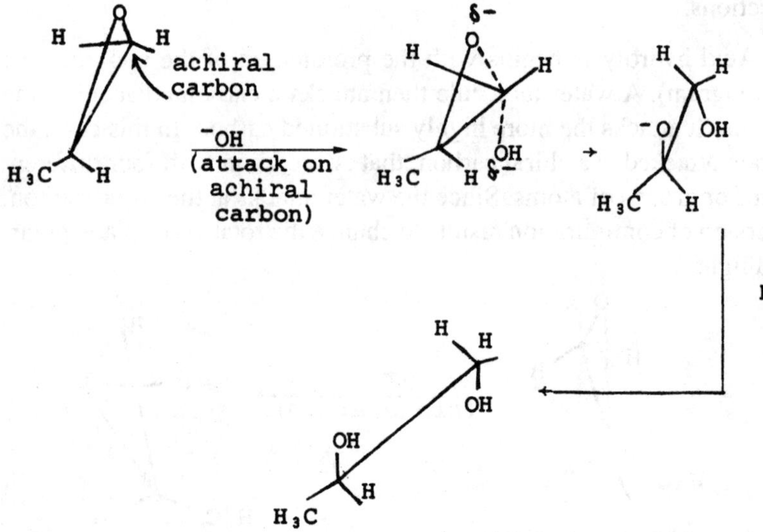

The products by base and acid hydrolysis are mirror images of each other. They rotate plane-polarized light in opposite directions.

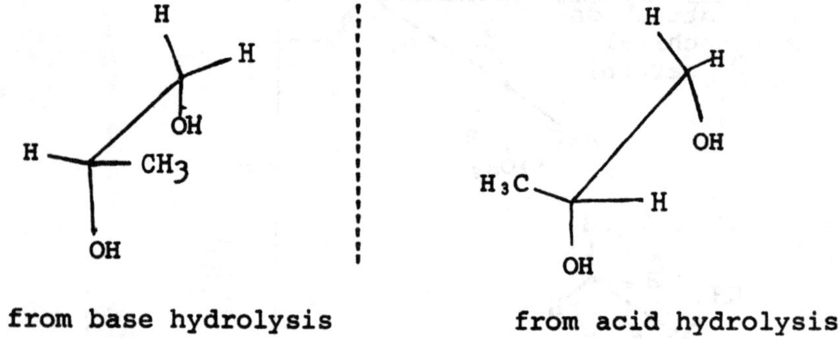

11.8 Reaction of Glycols

Oxidation of Ethanediol

$$\underset{\text{ethanediol}}{\begin{array}{c}CH_2OH\\|\\CH_2OH\end{array}} + \text{oxidation} \rightarrow \underset{\text{hydroxy ethanol}}{\begin{array}{c}CHO\\|\\CH_2OH\end{array}} \rightarrow \underset{\text{hydroxy ethanoic acid}}{\begin{array}{c}COOH\\|\\CH_2OH\end{array}} \rightarrow \underset{\text{oxalic acid}}{\begin{array}{c}CO{-}OH\\|\\CO{-}OH\end{array}}$$

Nitration of Ethanediol

$$HO{-}CH_2{-}CH_2{-}OH + 2HNO_3 \xrightarrow{H_2SO_4} O_2N{-}CH_2{-}CH_2{-}NO_2 + 2H_2O$$

1,2 –dinitroethane

Glycols undergo oxidative cleavage by periodic acid, yielding two carbonyl compounds:

$$\begin{array}{c}|\quad\ |\\ -C{-}C-\\ |\quad\ |\\ OH\ OH\end{array} + HIO_4 \rightarrow \;\rangle C{=}O + O{=}C\langle\; + H_2O + HIO_3$$

$$CH_3{-}CH(OH){-}C(OH)(CH_3){-}CH_3 + HIO_4 \rightarrow CH_3{-}C(H){=}O + O{=}C(CH_3){-}CH_3 + H_2O + HIO_3$$

Lead tetraacetate, $Pb(CH_3COO)_4$, cleaves glycols by oxidation.

Example

$$CH_3{-}CH(OH){-}C(OH)(CH_3){-}CH_2CH_3 + Pb(O{-}\overset{O}{\overset{\|}{C}}{-}CH_3)_4 \longrightarrow$$

$$CH_3{-}C(H){=}O + O{=}C(H_3C){-}CH_2{-}CH_3 + 2CH_3COOH + Pb(O_2CCH_3)_2$$

The Pinacol Rearrangement

Glycols may undergo rearrangement upon dehydration in acid. This reaction is known as the Pinacol Rearrangement. A carbonium ion is formed from the protonated diol. A 1,2 shift of an alkyl group occurs to produce the more stable carbonium ion. The driving force of the reaction is the conversion of the carbonium ion into the conjugate acid of the ketone. For example:

$$CH_3-C(CH_3)(OH)-C(CH_3)(OH)-CH_3 \xrightarrow{H^+} H_3C-C(CH_3)(H_2O^+)-C(CH_3)(OH)-CH_3 \xrightarrow{-H_2O} CH_3-\overset{+}{C}(CH_3)-C(CH_3)(OH)-CH_3$$

2, 3-dimethyl
2, 3-butanedisl
(pinacol)

$$\downarrow$$

$$CH_3-C(CH_3)_2-C(=\overset{+}{O}H)-CH_3 \xrightarrow{-H^+} CH_3-C(CH_3)_2-C(=O)-CH_3$$

t-Butyl methyl ketone
(pinacolone)

Problem Solving Examples:

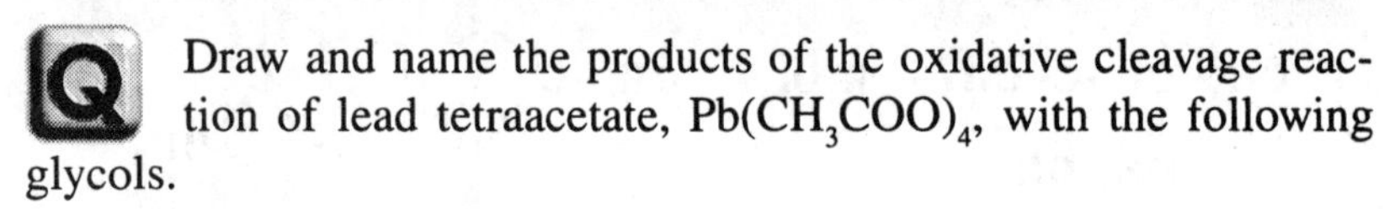

Q Draw and name the products of the oxidative cleavage reaction of lead tetraacetate, $Pb(CH_3COO)_4$, with the following glycols.

(a) $CH_3-CH(OH)-C(CH_3)(OH)-C_6H_5$

(b) 1-(1-hydroxyethyl)cyclohexanol: cyclohexane ring bearing OH and $-CH(OH)CH_3$ on the same carbon

A

(a)

$$CH_3-\underset{OH}{\overset{H}{C}}-\underset{OH}{\overset{CH_3}{C}}-C_6H_5 + Pb(CH_3COO)_4 \longrightarrow$$

$$CH_3-\overset{O}{\overset{\|}{C}} + CH_3-\overset{O}{\overset{\|}{C}}-C_6H_5 + \begin{matrix} 2\,CH_3COOH \\ Pb(O_2CCH_3)_2 \end{matrix}$$

(b)

$$C_6H_{10}(OH)-\underset{CH_3}{CH}(OH) + Pb(CH_3COO)_4 \longrightarrow$$

$$C_6H_{10}{=}O + CH_3\overset{O}{\overset{\|}{C}}H + \begin{matrix} 2CH_3COOH \\ Pb(O_2CCH_3)_2 \end{matrix}$$

Account for the products of the following reactions:

(a)

$$Ø-\underset{OH}{\overset{Ø}{C}}-\underset{OH}{\overset{Ø}{C}}-Ø \xrightarrow{H^+} Ø-\underset{Ø}{\overset{Ø}{C}}-\underset{O}{\underset{\|}{C}}-Ø$$

(b)

$$Ø-\underset{OH}{\overset{CH_3}{C}}-\underset{OH}{\overset{CH_3}{C}}-Ø \xrightarrow{H^+} Ø-\underset{Ø}{\overset{CH_3}{C}}-\underset{O}{\underset{\|}{C}}-CH_3$$

A In the case of unsymmetrical pinacols, two factors must be considered in determining the predominant product. The more stable carbonium ion is formed faster. The migration of an aryl group is preferred over a hydride migration, which in turn is favored over an alkyl migration.

(a)

$$Ø-\underset{OH}{\overset{Ø}{C}}-\underset{OH}{\overset{Ø}{C}}-Ø \xrightarrow{H^+} Ø-\underset{+OH_2}{\overset{Ø}{C}}-\underset{OH}{\overset{Ø}{C}}-Ø \longrightarrow Ø-\underset{+}{\overset{Ø}{C}}-\underset{:OH}{\overset{Ø}{C}}-Ø$$

$$\downarrow$$

$$Ø-\underset{Ø}{\overset{Ø}{C}}-\underset{O}{\underset{\|}{C}}-Ø \xleftarrow{-H^+} Ø-\underset{Ø}{\overset{Ø}{C}}-\underset{+OH}{\underset{\|}{C}}-Ø$$

(b)

```
  CH3 CH3                 CH3 CH3                   CH3 Ø
   |   |          H+       |   |                     |  |
Ø - C - C - Ø  ------>  Ø - C - C - Ø  ------>  Ø - C - C - CH3
   |   |                   |   |                     +  |
  OH  OH                 +OH2 OH                        OH
                                                         |
                                                         v
        CH3                                    CH3
         |                 - H+                 |
    Ø - C - C - CH3     <------           Ø - C - C - CH3
         |   ||                                 |   ||
         Ø   O                                  Ø  +OH
```

Quiz: Alcohols and Gylcols

1. The systematic (IUPAC) name of this structure is

```
     H   H   H   OH  H   H
     |   |   |   |   |   |
 H — C — C — C — C — C — C — H
     |   |   |   |   |   |
     H   H   H   H   H   H
```

(A) hexanol.
(B) 3-hydroxyhexane.
(C) 3-hexanol.
(D) 4-hexanol.
(E) isohexanol.

2. A secondary alcohol is formed through oxymercuration-demercuration of which one of the following?

(A) 1-hexene
(B) Hexane
(C) 2-methyl-1-bromopropene
(D) 2-2-dimethyl-propene
(E) 1-hexanal

3. Which one of the following reactions does not produce a high yield of alcohol?

(A) 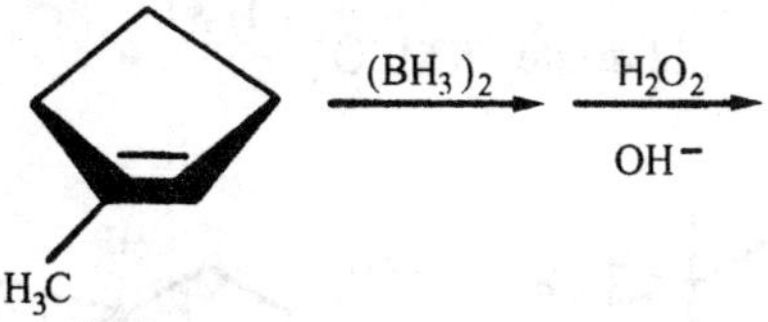 [cyclopentene ring with H_3C substituent] $\xrightarrow{(BH_3)_2}$ $\xrightarrow[OH^-]{H_2O_2}$

(B) 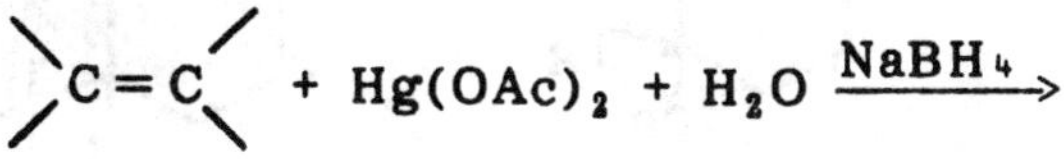 $>C=C<$ + $Hg(OAc)_2$ + H_2O $\xrightarrow{NaBH_4}$

(C) [benzene ring]-CH_2Cl $\xrightarrow{\text{aqueous NaOH}}$

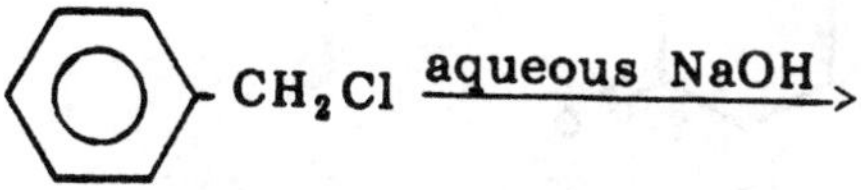

(D) 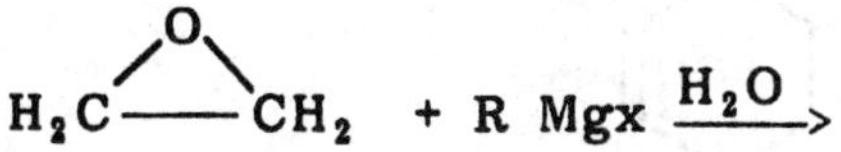 H_2C—CH_2 (bridged by O, epoxide) + R Mgx $\xrightarrow{H_2O}$

(E) 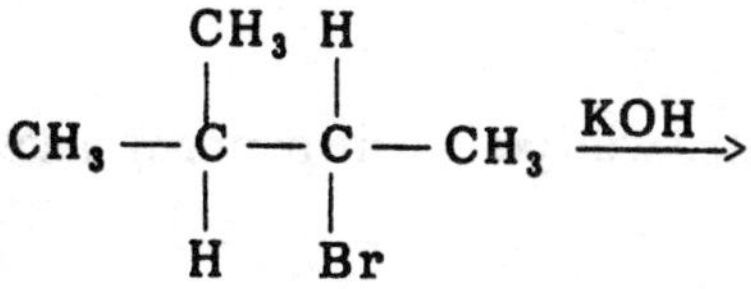 $CH_3-CH(CH_3)-CH(Br)-CH_3$ $\xrightarrow{KOH}$

4. Which one of the Grignard reactions could give rise to $CH_3CH_2CH(OH)CH_2CH_3$?

(A) Propanone and methyl Grignard

(B) Butyl Grignard and acetaldehyde

(C) Ethyl Grignard and propionaldehyde

(D) Methyl ethyl ketone and methyl Grignard

(E) Crotonaldehyde and ethyl Grignard

5. What is the product of the following reaction?

cyclohexene + cold dilute $KMnO_4$ $\xrightarrow{H_2O}$

(A) 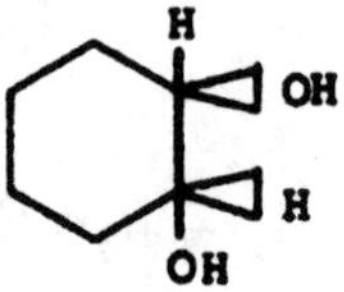 [structure: cyclohexane with H, OH / H, OH on adjacent carbons] (D) 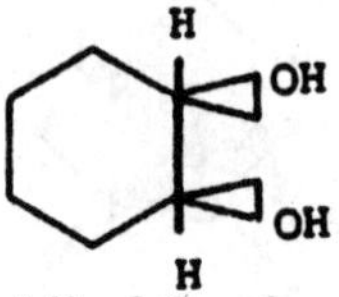 [structure: cyclohexane with H, OH / OH, H on adjacent carbons]

(B) 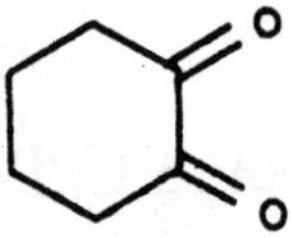 [structure: cyclohexane-1,2-dione] (E) All of the above.

(C)  [structure: cyclohexene]

6. Propene is treated with potassium permanganate. What is the major product?

(A) 1,2 ethane-diol (D) Ethanal

(B) 1,2 propanediol (E) Propan-2-ol

(C) Propanal

7. Which one of the following properties of alcohols decreases with an increase in C-content?

(A) Density (D) Boiling point

(B) Water solubility (E) None of the above.

(C) Refractive index

8. Alcohols are used for

(A) solvents.
(B) resins.
(C) antifreeze.
(D) preparation of esters.
(E) All of the above.

9. Glycols differ from alcohols in that they

(A) contain a carboxyl group.
(B) contain more than one hydroxyl group.
(C) contain a halide group.
(D) are prepared by oxymercuration-demercuration.
(E) participate in the Sandmeyer reaction.

10. The product of a pinacol rearrangement reaction is a(n)

(A) ketone.
(B) ester.
(C) glycol.
(D) carboxylic acid.
(E) None of the above.

ANSWER KEY

1.	(C)	6.	(B)
2.	(A)	7.	(B)
3.	(E)	8.	(E)
4.	(C)	9.	(B)
5.	(D)	10.	(A)

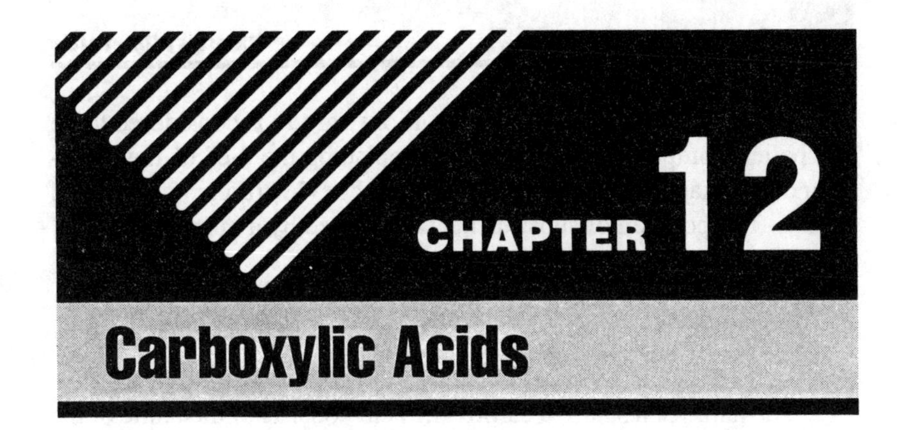

CHAPTER 12

Carboxylic Acids

Carboxylic acids contain a carboxyl group

$$-C\begin{matrix}\nearrow O \\ \searrow OH\end{matrix}$$

bonded to either an alkyl group (RCOOH) or an aryl group (ArCOOH).

HCOOH is formic acid (methanoic acid)

CH_3COOH is acetic acid (ethanoic acid)

C_6H_5—COOH is benzoic acid

12.1 Nomenclature (IUPAC System)

The longest chain carrying the carboxyl group is considered the parent structure and is named by replacing the "-e" ending of the corresponding alkane with "-oic acid."

$CH_3CH_2CH_2CH_2COOH$	Pentanoic acid
$CH_3CH\ \underset{\mid}{C}HCOOH$ $\quad\quad\ \ CH_3$	2-Methylbutanoic acid
C_6H_5—CH_2CH_2COOH	3-Phenylpropanoic acid
$CH_3CH{=}CHCOOH$	2-Butenoic acid

The longest chain carrying the carboxyl group is considered to be the parent compound. It is named by replacing the "-e" of the corresponding alkane with "-oic" acid. The carboxyl carbon is considered as C-1. The position of substituents is indicated by a number.

Example

$$CH_3CH_2\underset{\displaystyle CH_3}{\underset{|}{CH}}-CH_2COOH$$

3-Methyl pentanoic acid

The name of a salt of a carboxylic acid consists of the name of the cation followed by the name of the acid with the ending "-ic acid" changed to "-ate."

C_6H_5-COONa	$(CH_3COO)_2Ca$	$HCOONH_4$
Sodium benzoate	Calcium acetate	Ammonium formate

Problem Solving Examples:

What are carboxylic acids? Briefly discuss their properties and the system of naming.

Carboxylic acids are organic compounds that contain one or more carboxyl groups.

$$-C\begin{matrix}\nearrow\!\!\!\!\!\nearrow O \\ \searrow OH\end{matrix}$$

As the name implies, these compounds are acidic, in fact, they are the most important acidic organic compounds. This can be readily seen in the variety of substances that contain the carboxyl functional group. The most important examples of these substances are amino acids,

$$\left(R-\underset{\displaystyle NH_2}{\underset{|}{\overset{\displaystyle H}{\overset{|}{C}}}}-C\begin{matrix}\nearrow\!\!\!\!\!\nearrow O \\ \searrow OH\end{matrix} \right)$$

building blocks of proteins, which are required for animal function and structure.

The carboxyl group has a number of functional properties. They are as follows:

(a) The acidic proton can be released in the presence of a more basic compound, that is, a compound with higher affinity (pK_b) for proton.

(b) This acidic proton can participate in noncovalent hydrogen-bonding, as in protein tertiary and quarternary structures.

(c) The carbonyl oxygen can act as electron donor in hydrogen bonding.

(d) The carbonyl carbon is susceptible to nucleophilic attacks. It participates in various organic reactions, most notably in the amide bond formation in protein.

In the IUPAC system of naming, the longest chain carrying the carboxyl group is considered the parent structure, and is named by replacing the "-e" of the corresponding alkane with "-oic acid".

If there are other substitutions in the chain, they will be classified according to the carbon to which they are attached. The numbering of carbons begins with the carbon immediately attached to the carbonyl carbon, which is called the 2-carbon. The rest of the numbering will follow the Greek letters ß,Y,δ,ε...etc., which corresponds to 3, 4, 5, 6, . . .

$$\left(\begin{array}{l}\delta\ \ \gamma\ \ \beta\ \ \alpha \\ C-C-C-C-COOH\end{array}\right)$$

For example,

$$\overset{\beta}{CH_3}\overset{\alpha}{\underset{|}{CH}}C(=O)OH \quad (\alpha\text{-carbon bearing } OH)$$

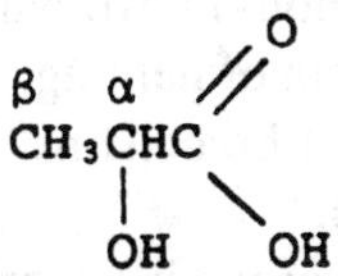

the longest chain contains three carbons; at the 2-carbon there is a hydroxy substitution. Thus, the IUPAC name is 2-hydroxypropionic acid.

There are also common names for some of the important acids. These are named according to the sources from which they are discovered. The most encountered ones are:

HCOOH	Formic acid

CH_3COOH	Acetic acid
$CH_3CH_2CH_2COOH$	Butyric acid

The 2-hydroxypropionic acid above is also called lactic acid.

Name the following structures:

(a) cyclohexane ring–COOH

(b) O_2N–(benzene ring)–COOH

(c) benzene ring with COOH, NO_2 (ortho) and NO_2 (para)

(d) $H_2N-\underset{\underset{CH(CH_3)_2}{|}}{\overset{\overset{H}{|}}{C}}-COOH$

(e) $HO_2C-CH_2-CH_2-CO_2H$

A Since the carboxyl group takes precedence in naming, all the above structures should have names endinq in -oic acid. According to the IUPAC system of naming, the compounds are named as derivatives of the longest carbon chains. Therefore, the nomenclature of the above compounds is:

(a) Since the carboxyl group is attached to a cyclohexane ring, the compound is called cyclohexanoic acid. This compound is also called cyclohexanecarboxylic acid.

(b) Substitutions in cyclic compounds are named according to the carbons to which they are attached. These carbons are classified according to their positions with respect to the main functional group. For example,

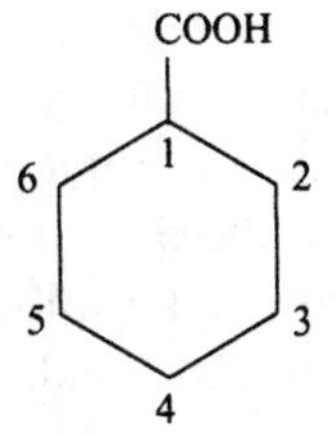

COOH

(benzene ring numbered 1–6, COOH at C1)

The C^2 and C^6 carbons are an equal distance from C^1 and the carboxyl carbon; they are called ortho-carbons. The C^3 and C^5 carbons are meta-carbons; the C^4 carbon is the para carbon. These three positions are abbreviated as the o, m, and p positions, respectively. Thus, the structure is called para-nitrobenzoic acid, or p-nitrobenzoic acid.

(c) When the cyclic compound has more than two substitutions, instead of using ortho, meta, or para, the carbon numbers are used.

Therefore, (ring with COOH at 1, NO_2 at 2, NO_2 at 4) is called 2,4-dinitrobenzoic acid.

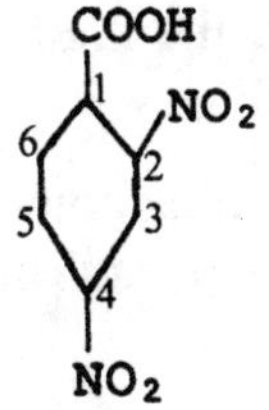

(d) The longest carbon chain contains four carbons. The substitutions are $-NH_2$ at carbon 2 and $-CH_3$ at carbon 3. Thus, the IUPAC name for the compound

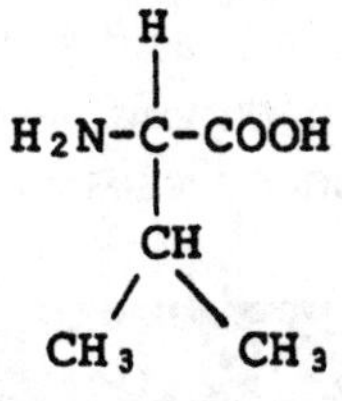

$$H_2N-\underset{\substack{|\\ CH(CH_3)_2}}{\overset{\substack{H\\|}}{C}}-COOH$$

is 2-amino,3-methyl butanoic acid.

However, this is also the structure for a natural, essential amino acid. The common name for it is valine.

(e) $HO_2C - CH_2 - CH_2 - CO_2H$ contains two carboxyl groups at two equivalent positions. Thus, naming can be from either carboxyl group, taking the other carboxyl as a substitution. The longest carbon chain containing one carboxyl group has three carbons; thus, it is called propionic acid. Since there is a carboxyl substitution at the 3-carbon, the full IUPAC name for the structure is 3-carboxylpropionic acid. The common name for the structure is succinic acid.

12.2 Physical Properties of Carboxylic Acids

A) Whether the attached group is aliphatic or aromatic, saturated or unsaturated, substituted or unsubstituted, the properties of the carboxyl group are essentially the same.

B) Carboxylic acids are polar molecules and can form hydrogen bonds with each other and with other kinds of molecules.

C) The first four members are miscible in water, the five-carbon acid is partially soluble, and the higher acids are insoluble. This is due to hydrogen bonding.

D) Carboxylic acids have high boiling points, which increase with carbon content, because pairs of molecules are held together by two hydrogen bonds.

E) Carboxylic acids are soluble in less polar solvents like ether, alcohol, benzene, etc.

F) The odors of lower aliphatic acids progress from the sharp irritating odors of formic acid and acetic acids to the distinctly unpleasant odors of butyric, valeric, and caproic acids. Higher acids have little odor because of their low volatility.

Problem Solving Examples:

Q (a) How many equivalents of base would be neutralized by one mole of phthalic acid? What is the neutralization equivalent of phthalic acid? (b) What is the relation between neutralization equivalent and the number of acidic hydrogens per molecule of acid? (c) What is the neutralization equivalent of 1,3,5-benzenetricarboxylic acid? Of mellitic acid, $C_6(COOH)_6$?

(a) Phthalic acid has the structure [structure: 1,2-benzenedicarboxylic acid, COOH / COOH]

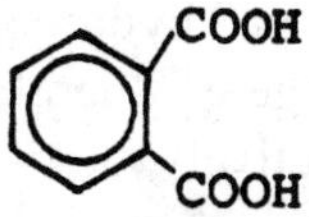

It is a dicarboxylic acid. Hence, for complete neutralization (the conversion by aqueous hydroxides of carboxylic acids into their salts), it will require two equivalents of base, one for each of the –COOH groups. For example,

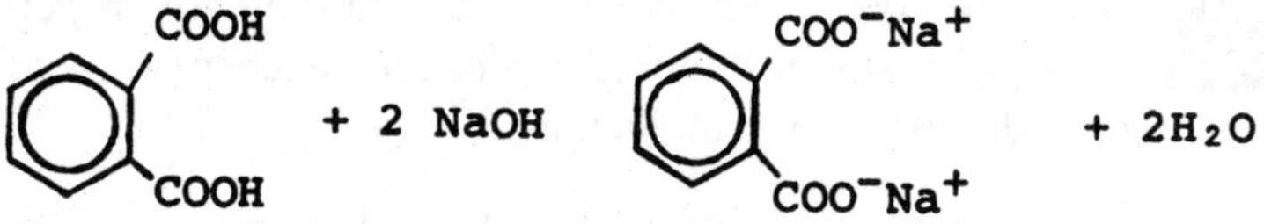

$$C_6H_4(COOH)_2 + 2\,NaOH \rightarrow C_6H_4(COO^-Na^+)_2 + 2H_2O$$

A useful way to identify and prove the structure of acids is by neutralization equivalent: the equivalent weight of the acid as determined by titration with standard base. Since phthalic acid takes two equivalents of base to be neutralized, its equivalent weight must be one-half of its molecular weight (166). Therefore, neutralization equivalent

$$= \frac{\text{molecular weight}}{2} = \frac{166}{2} = 83.$$

(b) It can be seen in (a) that the neutralization equivalent was obtained by dividing the molecular weight of the acid by the equivalents of base required for neutralization. The equivalents of base reflect, however, the number of acidic hydrogens (or COOHs) per molecule. Consequently,

$$\text{neutralization equivalent} = \frac{\text{molecular weight}}{\text{number of acidic H per molecule}}$$

(c) The structures of 1,3,5-benzenetricarboxylic acid and mellitic acid are, respectively,

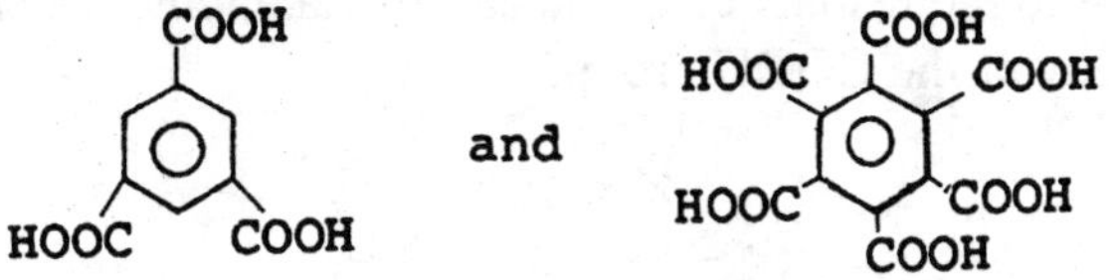

[structure: benzene ring with COOH, HOOC, COOH at 1,3,5 positions] and [structure: benzene ring with six COOH/HOOC groups]

The structures indicate that 1,3,5-benzenetricarboxylic acid has three acid hydrogens, whereas mellitic acid possesses six acid hydrogens. Therefore, the neutralization equivalents must be, respectively,

$$\frac{210}{3} = 70 \text{ and } \frac{342}{6} = 57.$$

Q How do you account for the fact that the α-hydrogens of an aldehyde (say, n-butyraldehyde) are much more acidic than any other hydrogens in the molecule?

$$\overset{\gamma}{CH_3}\overset{\beta}{CH_2}\overset{\alpha}{CH_2}\overset{t}{C} = O$$

n-butyraldehyde

A The acidity of the 2-hydrogens may be accounted for by the resulting resonance stability of the anion. The carbonyl group affects the acidity of the 2-hydrogens by helping to accommodate the negative charge of the anion. Ionization of the two-hydrogens yields a carbanion that is a resonance hybrid of two structures:

$$CH_3CH_2\overset{H}{\underset{\cdot\cdot}{\overset{|}{C}}}^{-}-\overset{H}{C}=\ddot{O}: \qquad CH_3CH_2\overset{H}{C}=\overset{H}{C}-:\ddot{O}:^{-}$$

These structures are equivalent to the carbanion.

$$CH_3CH_2\overset{H}{C}\text{=-=}\overset{H}{C}\text{=-=}O \quad (-)$$

Notice how the accommodation of this negative charge is similar in nature to the resonance of the carboxylate ions, which accounts for the acidity of the COOH group:

$$R-C{\Large(}^{O}_{O^-} \qquad R-C{\Large(}^{O^-}_{O}$$

is equivalent to

$$R-C\left\{\begin{matrix} O \\ O \end{matrix}\right\}^{-}$$

12.3 Preparation of Carboxylic Acids

Oxidation of Primary Alcohols

$$RCH_2OH \xrightarrow{KMnO_4} RCOOH$$

Example

$$CH_3CH_2CH(CH_3)CH_2OH \xrightarrow{KMnO_4} CH_3CH_2CH(CH_3)COOH$$

2-methyl-1-butanol → 2-methylbutanoic acid

Oxidation of Alkylbenzenes:

$$Ar-R \xrightarrow[K_2Cr_2O_7]{KMnO_4\text{, or}} Ar-COOH$$

$$O_2N-C_6H_4-CH_3 \xrightarrow[H_2SO_4\text{, heat}]{K_2Cr_2O_7} O_2N-C_6H_4-COOH$$

p-Nitrotoluene → p-Nitrobenzoic acid

Carbonation of Grignard Reagents:

$$\underset{(\text{or } ArX)}{RX} \xrightarrow{Mg} \underset{(\text{or } ArMgX)}{RMgX} \xrightarrow{CO_2} \underset{(\text{or } ArCOOMgX)}{RCOOMgX} \xrightarrow{H^+} \underset{(\text{or } ArCOOH)}{RCOOH}$$

$$Br-C_6H_4-CH(CH_3)C_2H_5 \xrightarrow{Mg} BrMg-C_6H_4-CH(CH_3)C_2H_5 \xrightarrow{CO_2} BrMgO_2C-C_6H_4-CH(CH_3)C_2H_5 \xrightarrow{H^+} HOOC-C_6H_4-CH(CH_3)C_2H_5$$

p-Bromo-sec-butyl benzene → p-sec-Butyl benzoic acid

Hydrolysis of Nitriles:

$$\underset{\text{or } Ar-C\equiv N}{R-C\equiv N} + H_2O \xrightarrow{\text{Acid or Base}} \underset{\text{or } Ar-COOH}{R-COOH} + NH_3$$

$$C_6H_5CH_2Cl \xrightarrow{NaCN} C_6H_5CH_2CN \xrightarrow[\text{reflux}]{70\%\ H_2SO_4} C_6H_5CH_2COOH + NH_3$$

Benzyl chloride → Phenyl-acetonitrile → Phenyl-acetic acid

Grignard Synthesis

$$R-MgX + O=C=O \rightarrow RCOO^-MgX^+ \xrightarrow{H^+} RCOOH + Mg^{++} + 2X^-$$

Nitrile Synthesis

$$R-X + CN^- \rightarrow R-C\equiv N + X^-$$

$$RC\equiv N + H_2O \xrightarrow{H^+} RCOOH + NH_4^+$$

$$RC\equiv N + H_2O \xrightarrow{OH^-} RCOO^- + NH_3$$

Friedel-Crafts acylation of aromatic hydrocarbons by acid anhydrides.

$$C_6H_6 + (H_2C-CO)_2O \xrightarrow{AlCl_3} C_6H_5-CO-CH_2-CH_2-COOH$$

Acid anhydride Ketocarboxylic acid

Problem Solving Examples:

Q What carboxylic acid can be prepared from p-bromotoluene: (a) by direct oxidation? (b) by free-radical chlorination followed by the nitrile synthesis?

A Carboxylic acids can be prepared in a variety of ways. One can employ direct oxidation with potassium permanganate, $KMnO_4$, or potassium dichromate, $K_2Cr_2O_7$, on primary alcohols and aldehydes. $KMnO_4$ may also be used to oxidize alkyl substituents on a benzene ring to the carboxyl group. Carboxylic acids can also be prepared by carbonation of Grignard reagents and hydrolysis of nitriles.

(a) Direct Oxidation:

p-bromotoluene is an alkylbenzene. Direct oxidation of the compound will change the methyl side chain to a carboxyl group. The oxidation reaction is illustrated on the next page.

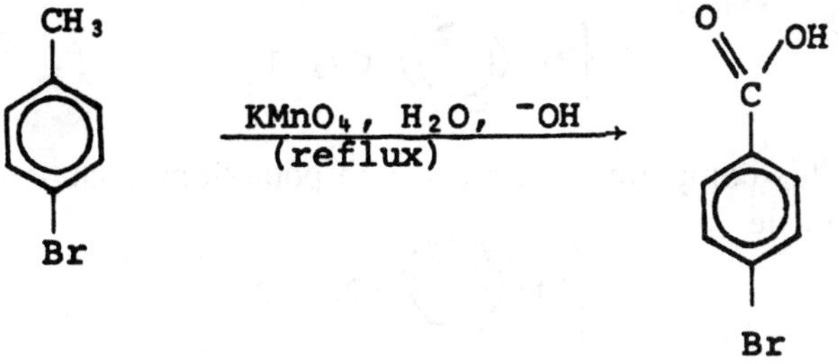

$$\text{p-bromotoluene} \xrightarrow[\text{(reflux)}]{KMnO_4,\ H_2O,\ ^{-}OH} \text{p-bromobenzoic acid}$$

p-bromotoluene p-bromobenzoic acid

(b) Free-Radical Chlorination Followed by Nitrile Synthesis: Alkylbenzenes are susceptible to free-radical halogenation. The halogen is added to the alkyl side chain such that the most stable radical intermediate is formed.

In general, a tertiary radical is more stable than a secondary radical, which is more stable than a primary radical. In equation form, the order of stability is

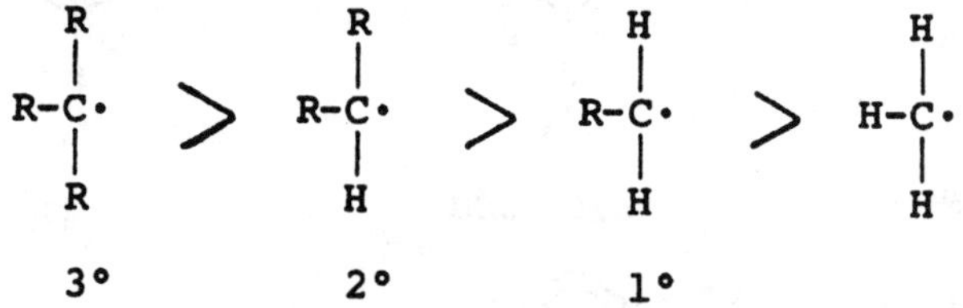

$$R_3C\cdot > R_2HC\cdot > RH_2C\cdot > H_3C\cdot$$

3° 2° 1°

where R stands for alkyl group. In the case of alkylbenzene, there is extra stability in the benzyl radical (that is, the radical that results from the abstraction of a hydrogen atom attached to the carbon joined directly to an aromatic ring). The stability of the benzyl radical may be accounted for by resonance stabilization. There exist five structures that contribute to the resonance hybrid. The hybrid may be pictured as follows:

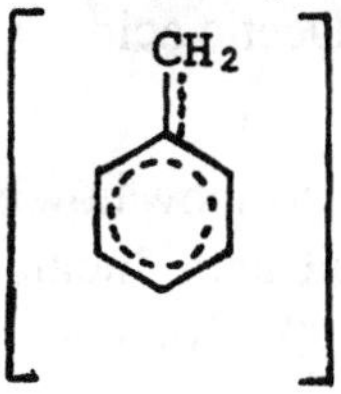

$$\left[C_6H_5 \text{---} CH_2 \right]$$

In this reaction, the free-radical chlorinated product is p-bromobenzyl chloride

$$Br-C_6H_4-CH_2Cl$$

This compound can react with potassium cyanide to form a nitrile

$$Br-C_6H_4-CH_2-C\equiv N$$

This reaction occurs because chlorine is a good leaving group, while $^{-}C\equiv N$ is a strong nucleophile. The nitrile formed can be hydrolized to give the carboxylic acid. The complete reaction is illustrated below.

$$Cl-Cl \xrightarrow{\text{light}} 2Cl\cdot$$

$$CH_3-C_6H_4-Br \xrightarrow{Cl\cdot} H_2C\cdot-C_6H_4-Br + HCl \xrightarrow{Cl\cdot} H_2CCl-C_6H_4-Br \xrightarrow{K^+C\equiv N^-}$$

$$H_2CCN-C_6H_4-Br \longrightarrow H_2C\,COOH-C_6H_4-Br$$

p-bromo phenylacetic acid

The carboxylic acid formed by free-radical chlorination of p-bromotoluene, followed by nitrile synthesis and hydrolysis, is p-bromophenylacetic acid.

Q Write equations to show how each of the following can be converted to benzoic acid. Include all reagents and conditions.

(a) Benzyl alcohol (b) Toluene (c) Benzaldehyde

A Carboxylic acids can be prepared from alcohols, ketones, and aldehydes by oxidation. Primary alcohols may be readily oxidized to carboxylic acids by chromic acid. Chromic acid, $H_2Cr_2O_7$, is a strong oxidizing agent prepared by adding sulfuric acid to sodium dichromate. It readily oxidizes primary alcohols to aldehydes, and the aldehydes are, in turn, rapidly oxidized to carboxylic acids.

$$R{-}CH_2OH \xrightarrow{H_2Cr_2O_7} R{-}\underset{\underset{O}{\|}}{C}{-}H \xrightarrow{H_2Cr_2O_7} R{-}\underset{\underset{O}{\|}}{C}{-}OH$$

Aldehydes are very easily oxidized to carboxylic acids—even atmospheric oxygen at room temperature will gradually convert aldehydes to acids. For synthetic purposes, chromic acid is a satisfactory reagent for this reaction.

$$R - CHO \xrightarrow{H_2Cr_2O_7} R{-}COOH$$

When an alkyl substituent is attached to a benzene ring, the substituent may be oxidized by treatment with hot potassium permanganate. Regardless of the length of the alkyl chain, only one carbon remains attached to the ring in the final product, and this carbon is in the form of a carboxyl group. Any other carbons that were originally present in the chain are converted to carbon dioxide.

$$C_6H_5{-}R \xrightarrow[\text{heat}]{KMnO_4} C_6H_5{-}COOH$$

(a) For benzyl alcohol $(C_6H_5{-}CH_2OH)$ which is a primary alcohol, can be oxidized using $H_2Cr_2O_7$ to benzoic acid as follows:

$$C_6H_5{-}CH_2OH \xrightarrow{H_2Cr_2O_7} C_6H_5{-}COOH$$

(b) Toluene is an alkyl substituted benzene ring. It undergoes oxidation with hot $KMnO_4$ to give benzoic acid as follows:

$$C_6H_5CH_3 \xrightarrow{KMnO_4} C_6H_5COOH$$

(c) Benzaldehyde, which has the structure

$$C_6H_5-C(H)=O$$

undergoes oxidation with $H_2Cr_2O_7$ to give benzoic acid. Thus,

$$C_6H_5CHO \xrightarrow{H_2Cr_2O_7} C_6H_5COOH$$

12.4 Reactions of Carboxylic Acids

Acidity Salt Formation

$$RCOOH \rightleftharpoons RCOO^- + H^+$$

$$\underset{\text{Acetic acid}}{2CH_3COOH} + Zn \longrightarrow \underset{\text{Zinc acetate}}{(CH_3COO^-)_2\,Zn^{++}} + H_2$$

$$\underset{\text{Benzoic acid}}{C_6H_5COOH} + NaHCO_3 \longrightarrow \underset{\text{Sodium benzoate}}{C_6H_5COO^-Na^+} + CO_2 + H_2O$$

Conversion into Functional Derivatives

$$R-C(=O)-OH \rightarrow R-C(=O)-Z \quad \text{(where } Z = -Cl, -OR', -NH_2\text{)}$$

A) Conversion to acid chlorides:

$$R{-}C(=O)OH + \left\{\begin{array}{l} SOCl_2 \\ PCl_3 \\ PCl_5 \end{array}\right\} \longrightarrow R{-}C(=O)Cl$$

Acid chloride

$$C_6H_5{-}COOH + PCl_5 \xrightarrow{100^\circ C} C_6H_5{-}COCl + POCl_3 + HCl$$

Benzoic acid → Benzoyl chloride

$$3CH_3COOH + PCl_3 \xrightarrow{50^\circ C} 3CH_3COCl + H_3PO_3$$

Acetic acid → Acetyl chloride

B) Conversion into esters:

$$R{-}C(=O)OH + R'OH \underset{}{\overset{H^+}{\rightleftharpoons}} R{-}C(=O)OR' + H_2O$$

ester

$$R{-}C(=O)OH \xrightarrow{SOCl_2} R{-}C(=O)Cl \xrightarrow{R'OH} R{-}C(=O)OR'$$

acid chloride → ester

$$C_6H_5{-}COOH + CH_3OH \overset{H^+}{\rightleftharpoons} C_6H_5{-}COOCH_3 + H_2O$$

Benzoic acid, Methanol → Methyl benzoate

$$CH_3COOH + C_6H_5{-}CH_2OH \overset{H^+}{\rightleftharpoons} CH_3COOCH_2{-}C_6H_5 + H_2O$$

Acetic acid, Benzyl alcohol → Benzyl acetate

C) Conversion to amides:

$$R{-}C(=O){-}OH \xrightarrow{SOCl_2} R{-}C(=O){-}Cl \xrightarrow{NH_3} R{-}C(=O){-}NH_2$$

acid chloride — amide

Example

$$C_6H_5CH_2COOH \xrightarrow{SOCl_2} C_6H_5CH_2COCl \xrightarrow{NH_3} C_6H_5CH_2CONH_2$$

phenylacetic acid — phenylacetyl chloride — phenylacetamide

Reduction

$$RCOOH \xrightarrow{LiAlH_4} RCH_2OH$$

alcohol (1°)

Example

$$4(CH_3)_3CCOOH + 3LiAlH_4 \xrightarrow{ether} [(CH_3)_3CCH_2O]_4AlLi + 2LiAlO_2 + 4H_2$$

Trimethylacetic acid

$$\xrightarrow{H^+} (CH_3)_3CCH_2OH$$

2,2-dimethyl-1-propanol

Substitution in alkyl or aryl group:

a) Alpha-halogenation of aliphatic acids

(Hell-Volhard-Zelinsky reaction)

$$R{-}CH_2COOH + X_2 \xrightarrow{P} RCH(X)COOH + HX \quad (X = Cl, Br)$$

α-haloacid

Example

$$CH_3COOH \xrightarrow{Cl_2,P} ClCH_2COOH \xrightarrow{Cl_2,P}$$

acetic acid — chloroacetic acid

$$\rightarrow Cl_2CHCOOH \rightarrow Cl_3CCOOH$$

dichloroacetic acid — trichloroacetic acid

$$CH_3\underset{\displaystyle CH_3}{\overset{|}{C}}HCH_2COOH \xrightarrow{Br_2,P} CH_3-\underset{}{\overset{\displaystyle CH_3}{\overset{|}{C}}}H-\underset{\displaystyle Br}{\underset{|}{C}}H-COOH$$

isovaleric acid — α-bromoisovaleric acid

b) Ring substitution in aromatic acids

–COOH: deactivates and directs meta in electrophilic substitution.

$$C_6H_5COOH \xrightarrow[\text{heat}]{HNO_3,\ H_2SO_4} m\text{-}NO_2C_6H_4COOH$$

Benzoic acid — m-Nitro-benzoic acid

Problem Solving Examples:

Explain the mechanism involved in the nucleophilic substitution reaction on carboxylic acids in esterification.

A The most important example of nucleophilic substitution on carboxylic acids is esterification, in which carboxylic acids react with alcohols to give esters.

$$R-\underset{\displaystyle O}{\underset{\|}{C}}-OH \quad + \quad R'-OH \quad \underset{}{\overset{H^+}{\rightleftharpoons}} \quad R-\underset{\displaystyle O}{\underset{\|}{C}}-O-R' \quad + \quad H_2O$$

Carboxylic acid — Alcohol — Ester

The esterification reaction requires the presence of a strong acid catalyst. The first step of the reaction is the protonation of the carboxyl group at the double-bonded oxygen.

$$R\text{-}C(=O)\text{-}OH + H^+ \rightleftharpoons \left\{ R\text{-}C(=\overset{+}{O}H)\text{-}OH \longleftrightarrow R\text{-}C^+(\text{-}OH)\text{-}OH \right\} \equiv \left[R\text{-}\overset{\delta+}{C}(\text{-}OH)\text{=}\overset{\delta+}{O}\text{-}H \right]^+$$

The protonated carboxyl group is readily attacked by nucleophiles (in this case R′ – OH) because the carboxyl carbon now has significant positive character.

$$\left[R\text{-}\overset{\delta+}{C}(\text{-}OH)\text{=}\overset{\delta+}{O}\text{-}H \right]^+ + R\text{—}\ddot{O}H \rightleftharpoons R\text{-}C(\text{-}\overset{+}{O}H\text{-}R')(\text{-}OH)\text{-}O\text{-}H$$

(Adduct)

The adduct formed by this attack undergoes a hydrogen ion transfer and an elimination to give the final product.

$$\underset{\text{Adduct}}{R'\text{-}\overset{+}{O}H\text{-}C(R)(OH)\text{-}O\text{-}H} \xrightleftharpoons{H^+\ \text{transfer}} R'\text{-}O\text{-}C(R)(\text{-}\overset{+}{O}H_2)\text{-}O\text{-}H \xrightleftharpoons{\text{elimination}} \underset{\text{Ester}}{R'\text{-}O\text{-}C(R)\text{=}O}$$

$+H_2O + H^+$

To recapitulate, the steps of the mechanism are:

(1) Protonation of the carboxylic acid:

$$R{-}C({=}O){-}OH + H^+ \rightleftharpoons R{-}C({=}\overset{+}{O}{-}H){-}OH$$

(2) Attack by R' – OH:

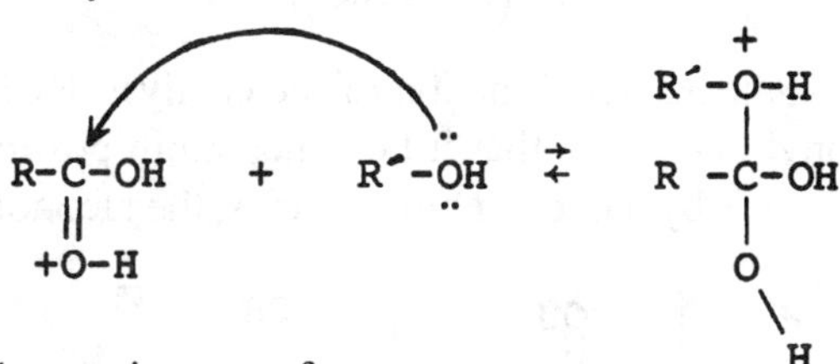

(3) Hydrogen ion transfer:

$$R'{-}\overset{+}{O}H{-}C(R)(OH){-}OH \rightleftharpoons R'{-}O{-}C(R)(OH){-}\overset{+}{O}H_2$$

(4) Elimination of water and H^+:

$$R'{-}O{-}C(R)(OH){-}\overset{+}{O}H_2 \rightleftharpoons R'{-}O{-}C(R){=}O + H_2O + H^+$$

Since every step of the esterification reaction is reversible, the reaction is usually driven to completion by employing either the alcohol or the carboxylic acid in great excess. For example, benzoic acid may be almost completely converted to the corresponding ethyl ester, ethyl benzoate, by using ethanol as the solvent.

$$C_6H_5{-}COOH + CH_3CH_2OH \underset{}{\overset{H^+}{\rightleftharpoons}} C_6H_5{-}COOCH_2CH_3 + H_2O$$

Benzoic acid   Ethanol (excess)   Ethyl benzoate

Predict the outcome of an attempted esterification of acetic acid with t-butyl alcohol in the presence of dry HCl.

An examination of the mechanism for acid-catalyzed formation of esters

$$\left(R-\overset{O}{\overset{\|}{C}}-OR'\right)$$

will aid in the solution. The function of the catalyst, HCl, is to protonate the carbonyl carbon so that it becomes more positive and more susceptible to attack by a nucleophilic species, the alcohol in this case.

$$\underset{\text{(Carboxylic acid)}}{R-\underset{\underset{O}{\|}}{C}-OH} \xrightleftharpoons{H^+} \left[R-\overset{^+OH}{\overset{\|}{C}}-OH \leftrightarrow R-\underset{+}{\overset{OH}{\overset{|}{C}}}-OH \right] \underset{\text{(Alcohol)}}{\xrightleftharpoons{R'OH}} \left[R-\underset{\underset{\underset{+}{R'OH}}{|}}{\overset{OH}{\overset{|}{C}}}-OH \right]$$

[1]

The product [1] formed can rapidly and reversibly form a second intermediate [2] by proton transfer as shown:

$$\underset{[1]}{\left[R-\underset{\underset{\underset{+}{R'OH}}{|}}{\overset{OH}{\overset{|}{C}}}-OH \right]} \rightleftarrows \underset{[2]}{\left[R-\underset{\underset{R'O}{|}}{\overset{^+OH_2}{\overset{|}{C}}}-OH \right]}$$

If [2] loses a molecule and it is followed by transfer of a proton, it gives the ester [3].

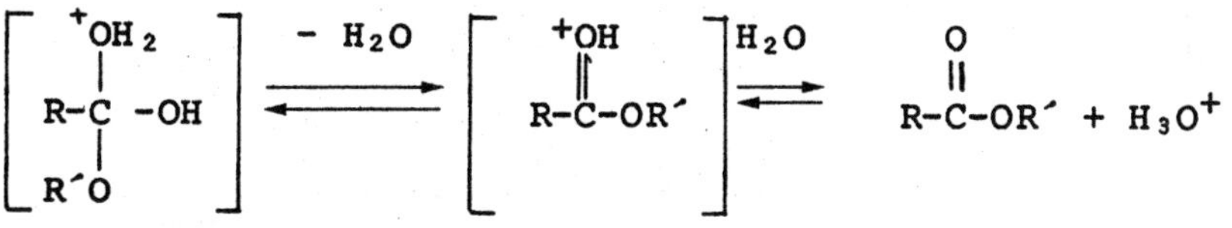

$$\left[R{-}\overset{\overset{+}{O}H_2}{\underset{R'O}{C}}{-}OH \right] \underset{}{\overset{-H_2O}{\rightleftharpoons}} \left[R{-}\overset{\overset{+}{O}H}{\overset{\|}{C}}{-}OR' \right] \overset{H_2O}{\rightleftharpoons} R{-}\overset{O}{\overset{\|}{C}}{-}OR' + H_3O^+$$

[2] [3]

With this in mind, one might think the outcome of the reaction between acetic acid (CH_3COOH) and t-butyl alcohol in dry HCl would be t-butyl acetate

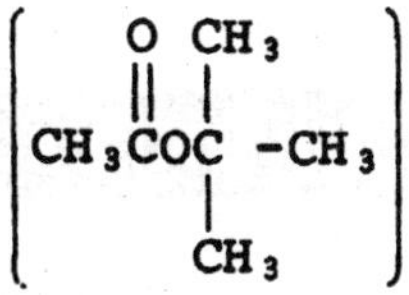

$$\left(CH_3\overset{O}{\overset{\|}{C}}O\overset{CH_3}{\underset{CH_3}{C}}{-}CH_3 \right)$$

Such thinking is incorrect. The reason stems from the fact that if the alcohol is particularly bulky, the reaction will usually not proceed satisfactorily, since the intermediate [1] or [2] is rendered unstable by crowding of the substituent groups. (Bulky groups in the esterifying acid also hinder the reaction.) Hence, no t-butyl acetate is formed due to the bulkiness of the t-butyl alcohol. What actually occurs is that the alcohol and the hydrogen halide (HCl) react to form the alkyl halide as shown:

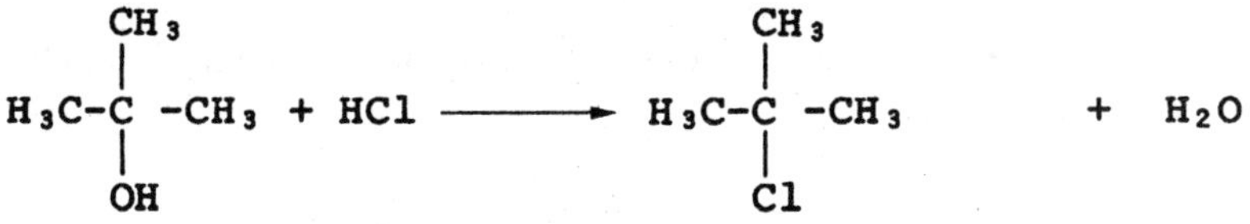

$$H_3C{-}\overset{CH_3}{\underset{OH}{C}}{-}CH_3 + HCl \longrightarrow H_3C{-}\overset{CH_3}{\underset{Cl}{C}}{-}CH_3 + H_2O$$

(t-butyl chloride)

So, the product is t-butyl chloride.

12.5 Acidity of Carboxylic Acids

The acidity of carboxylic acid is due to the powerful resonance-stabilization of its anion. This stabilization and resulting acidity are possible only because of the presence of the carbonyl group. The stabilization can be seen in the figure on the next page.

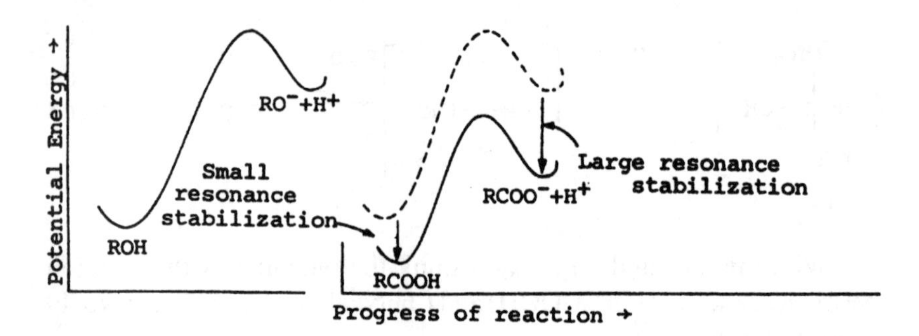

Problem Solving Example:

Q List the following carboxylic acids in order of increasing acidity. Explain your answer in terms of the corresponding carboxylate anion.

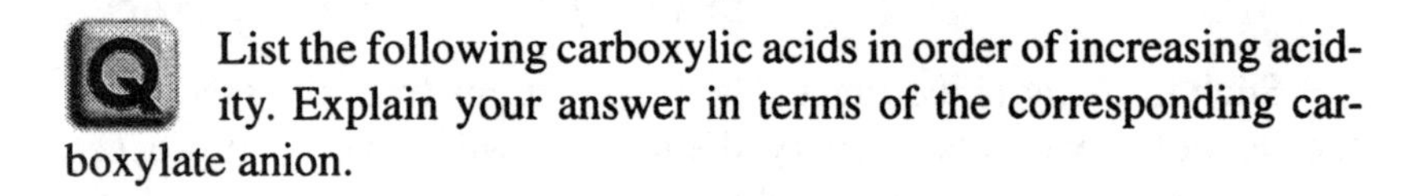

$$Cl-\underset{H}{\overset{Cl}{C}}-\underset{O}{\overset{}{C}}-OH, \quad CH_3CH_2-\underset{O}{C}-OH, \quad \underset{Cl}{CH_2}-\underset{O}{C}-OH$$

A

$$CH_3CH_2-\underset{O}{C}-OH < \underset{Cl}{CH_2}-\underset{O}{C}OH < Cl_2-CH-\underset{O}{C}-OH$$

Carboxylic acids are acidic in aqueous solutions. The ionization proceeds as follows:

$$R-\underset{O}{C}-OH \rightleftharpoons R-\underset{O}{C}-O^- + H^+$$

The carboxylate anion, $R-\underset{O}{C}-O^-$, is capable of resonance delocalization of charge.

$$R-\underset{O}{C}-O^- \rightleftharpoons R-\underset{O^-}{C}=O$$

Electron-releasing substituents destabilize the anion and decrease acidity.

Electron-withdrawing groups stabilize the anion.

12.6 Structure of Carboxylate Ions

A carboxylate ion, according to the resonance theory, is a hybrid of two structures which, being of equal stability, contribute equally. Carbon is joined to each oxygen by a "one and one-half" bond.

Ionization of Carboxylic Acids: Acidity Constants

Carboxylic acids are acidic in aqueous solutions. The ionization of carboxylic acids proceeds as follows:

$$\underset{\text{carboxylic acid}}{R-\overset{\overset{\displaystyle O}{\|}}{C}-OH} \rightleftharpoons \underset{\text{carboxylate anion}}{R-\overset{\overset{\displaystyle O}{\|}}{C}-O^-} + H^+$$

The ionization of a carboxylic acid produces a carboxylate anion R–COO⁻, which is capable of resonance delocalization of charge.

$$R-\overset{\overset{\displaystyle O}{\|}}{C}-OH \rightleftharpoons \left[\begin{array}{c} R-\overset{\overset{\displaystyle O}{\|}}{C}-O^- \\ \\ R-\underset{\underset{\displaystyle O^-}{|}}{C}=O \end{array}\right] + H^+$$

Resonance delocalization

Electron-withdrawing substituents stabilize the anion and increase acidity.

Electron-releasing substituents destabilize the anion and decrease acidity.

The electron-withdrawing halogens strengthen acids. For example, in order of increasing acidity:

Acetic Acid < Chloroacetic Acid < Dichloroacetic Acid < Trichloro–acetic Acid

Acidity Constant

$$RCOOH + H_2O \rightleftharpoons RCOO^- + H_3O^+$$

$$K_a = \frac{[RCOO^-][H_3O^+]}{[RCOOH]}$$

K_a is the acidity constant.

K_a, which is a characteristic of every carboxylic acid, indicates the strength of the acid. As the value K_a increases, the extent of ionization and the strength of the acid increase.

Relative acidities $RCOOH > HOH > ROH > HC \equiv CH > NH_3 > RH$

Relative basicities $RCOO^- < HO^- < RO^- < HC \equiv C^- < NH_2^- < R^-$

CHAPTER 13

Carboxylic Acid Derivatives

Carboxylic acid derivatives are compounds in which the hydroxyl group has been replaced by –Cl, –OOCR, $-NH_2$, or –OR′. These derivatives are called acid chlorides, anhydrides, amides, and esters, respectively.

Problem Solving Examples:

What are the carboxylic acid derivatives? Discuss their nomenclature. Give examples.

A The carboxyl group of organic acids may exist in several modified forms. A different entity may replace the hydroxyl portion —such things are called carboxylic acid derivatives.

When the hydroxyl is replaced by chlorine, an acyl chloride is obtained:

$$\mathrm{R{-}\underset{\underset{O}{\|}}{C}{-}Cl}$$

The nomenclature of an acylchloride is determined from the corresponding acid. The ending "-ic acid" is replaced by the ending "-yl chloride." For example, propanoic acid (CH_3CH_2COOH) would become propanoyl chloride.

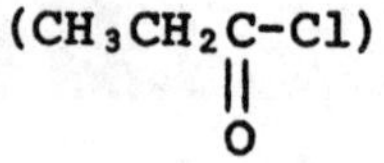

$$(CH_3CH_2\underset{\underset{O}{\|}}{C}-Cl)$$

These substances can also be named, respectively, propionic acid and propionyl chloride.

If an alkoxyl replaces the hydroxyl, an ester is formed:

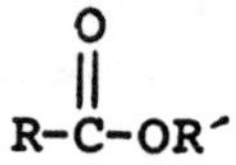

$$R-\overset{\overset{O}{\|}}{C}-OR'$$

The name of an ester includes two words: The first derives from the alcohol; the second comes from the name of the carboxylic acid by replacing the ending "-ic" with the ending "-ate." For example,

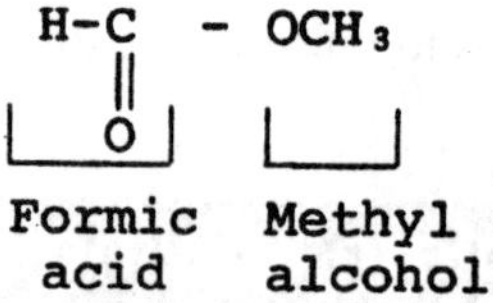

$$\underbrace{H-\underset{\underset{O}{\|}}{C}}_{\text{Formic acid}} - \underbrace{OCH_3}_{\text{Methyl alcohol}}$$

the ester is derived from methyl alcohol and formic acid. Therefore, its name becomes methyl formate. Similarly, the ester below

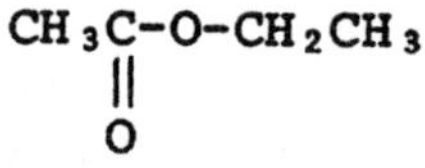

$$CH_3\underset{\underset{O}{\|}}{C}-O-CH_2CH_3$$

is named ethyl acetate (ethyl ethanoate).

Carboxamides or amides come about when the hydroxyl group is replaced by an amino group. If the amino group has attached to it one or two alkyl or aryl groups, the amide is referred to as N-substituted or N,N-disubstituted.

$$R{-}\overset{\displaystyle O}{\overset{\|}{C}}{-}NH_2 \qquad R{-}\overset{\displaystyle O}{\overset{\|}{C}}{-}N(R')(H) \qquad R{-}\underset{\displaystyle O}{\underset{\|}{C}}{-}N(R')(R'')$$

(Simple amide) (N-Substituted amide) (N,N-Disubstituted amide)

The simple amides are named from the corresponding carboxylic acid by dropping the ending "-oic acid" or "-ic acid" and substituting the ending "-amide." For example, acetic acid or ethanoic acid

$$(CH_3\underset{\displaystyle O}{\underset{\|}{C}}{-}OH)$$

becomes acetamide or ethanamide

$$(CH_3\underset{\displaystyle O}{\underset{\|}{C}}{-}NH_2)$$

The N-substituted amides are named by indicating the N-substituents before the name of the parent amide. For example,

$$CH_3\underset{\displaystyle O}{\underset{\|}{C}}{-}NH{-}CH_3$$

would be N-methylacetamide.

Carboxylic acid anhydrides may be viewed as compounds derived from carboxylic acids by the loss of water:

$$R{-}\underset{\displaystyle O}{\underset{\|}{C}}{-}O{-}\underset{\displaystyle O}{\underset{\|}{C}}{-}R$$

In naming anhydrides, the word "acid" is dropped from the component carboxylic acid and replaced with "anhydride." For example,

$$CH_3\underset{\displaystyle O}{\underset{\|}{C}}{-}O{-}\underset{\displaystyle O}{\underset{\|}{C}}CH_3$$

would be called acetic anhydride or ethanoic anhydride.

13.1 Acid Chlorides

Nomenclature (IUPAC System)

When naming acid chlorides, the ending "-ic acid" in the carboxylic acid is replaced by the ending "-yl chloride."

$R-C(=O)Cl$ Acid chloride

$CH_3-C(=O)Cl$ Ethanoyl chloride (Acetyl chloride)

$C_6H_5-C(=O)Cl$ Benzoyl chloride

$CH_3CH_2C(=O)-Cl$ Propanoyl chloride

$CH_3CH(CH_3)CH_2CH_2C(=O)-Cl$ 4-Methyl pentanoyl chloride

$m-O_2NC_6H_4C(=O)-Cl$ m-Nitrobenzoyl chloride

$CH_3CH=CHCOCl$ 2-Butenoyl chloride

$C_6H_{11}-COCl$ Cyclohexane carbonyl chloride

Physical Properties of Acid Derivatives

The presence of the C = O group makes the acid derivatives polar compounds. Acid chlorides, anhydrides, and esters have boiling points that are about the same as those of aldehydes and ketones of comparable molecular weight. The acid derivatives are soluble in the usual organic solvents.

Acid chlorides have sharp, irritating odors, partly because they are readily hydrolyzed to HCl and carboxylic acid. It is because of this that they must be protected from moisture.

Acid chlorides are used as intermediates in synthesis. They are fairly reactive compounds and are usually liquids.

Preparation of Acid Chlorides

$$R-\overset{O}{\overset{\|}{C}}-OH \xrightarrow[\text{or } SOCl_2]{PCl_5 \text{ or } PCl_3} R-\overset{O}{\overset{\|}{C}}-Cl$$

carboxylic acid → acid chloride

$$C_6H_5-COOH + SOCl_2 \xrightarrow{\text{reflux}} C_6H_5-COCl + SO_2 + HCl$$

Benzoic acid → Benzoyl chloride

$$3,5\text{-}(O_2N)_2C_6H_3-COOH + PCl_5 \xrightarrow{\text{heat}} 3,5\text{-}(O_2N)_2C_6H_3-COCl + POCl_3 + HCl$$

3,5-Dinitrobenzoic acid → 3,5-Dinitrobenzoyl chloride

$$3\,CH_3COOH + PCl_3 \xrightarrow{50°C} 3CH_3COCl + H_3PO_3$$

Acetic acid → Acetyl chloride

Reactions of Acid Chlorides

A) Conversion into acids and derivatives

$$R-C(=O)Cl + HZ \rightarrow R-C(=O)Z + HCl$$

a) Conversion into acids. Hydrolysis.

$$RCOCl + H_2O \rightarrow RCOOH + HCl$$

an acid

$$C_6H_5-COCl + H_2O \rightarrow C_6H_5-COOH + HCl$$

Benzoyl chloride → Benzoic acid

b) Conversion into amides. Ammonolysis.

$$RCOCl + 2NH_3 \longrightarrow RCONH_2 + NH_4Cl$$
An amide

$$C_6H_5COCl + 2NH_3 \longrightarrow C_6H_5CONH_2 + NH_4Cl$$
Benzoyl chloride — Benzamide

$$R{-}\underset{\underset{O}{\|}}{C}{-}Cl + R'{-}NH_2 \rightarrow R{-}\underset{\underset{O}{\|}}{C}{-}NH{-}R' + HCl$$

primary amine — N-substituted carboxamide

$$R{-}\underset{\underset{O}{\|}}{C}{-}Cl + R'{-}NH{-}R'' \rightarrow R{-}\underset{\underset{O}{\|}}{C}{-}N\begin{matrix} R' \\ R'' \end{matrix} + HCl$$

secondary amine — N,N-disubstituted carboxamide

c) Conversion into esters. Esterification.

$$RCOCl + R'OH \longrightarrow RCOOR' + HCl$$
An ester

$$C_6H_5COCl + C_2H_5OH \longrightarrow C_6H_5COOC_2H_5 + HCl$$
Benzoyl chloride — Ethyl alcohol — Ethyl benzoate

B) Formation of ketones. Friedel-Crafts acylation.

$$R-C(=O)Cl + ArH \xrightarrow[\text{or other Lewis acid}]{AlCl_3} R-C(=O)-Ar + HCl$$

A ketone

$$m\text{-}O_2N\text{-}C_6H_4\text{-}COCl + C_6H_6 \xrightarrow{AlCl_3} m\text{-}O_2N\text{-}C_6H_4\text{-}C(=O)\text{-}C_6H_5$$

m-Nitrobenzoyl chloride, Benzene → m-Nitrobenzophenone

C) Formation of ketones. Reaction with organocadmium compounds.

$$R'MgX \xrightarrow{CdCl_2} R'_2Cd$$

R'_2Cd + RCOCl or ArCOCl → $R-C(=O)-R'$ or $Ar-C(=O)-R'$

$$2O_2N\text{-}C_6H_4\text{-}COCl + (CH_3)_2Cd \longrightarrow 2O_2N\text{-}C_6H_4\text{-}C(=O)\text{-}CH_3 + CdCl_2$$

p-Nitrobenzoyl chlorine, Dimethyl cadmium → p-Nitroacetophenone (Methyl p-nitrophenyl ketone)

D) Formation of aldehydes by reduction.

a) $$RCOCl \text{ or } ArCOCl \xrightarrow{LiAlH(OBu\text{-}t)_3} RCHO \text{ or } ArCHO$$

Aldehyde

$$O_2N\text{-}C_6H_4\text{-}COCl \xrightarrow{LiAlH(OBu\text{-}t)_3} O_2N\text{-}C_6H_4\text{-}CHO$$

p-Nitrobenzoyl chloride → p-Nitrobenzaldehyde

b) Rosenmund reduction of acid chlorides.

$$R-C(=O)-Cl + H_2 \xrightarrow{Pd/BaSO_4} RCHO + HCl$$

aldehyde

E) Reduction to alcohols.

$$2CH_3COCl + LiAlH_4 \rightarrow LiAlCl_2(OCH_2CH_3)_2 \xrightarrow{H^+} 2CH_3CH_2OH$$

Problem Solving Examples:

Suggest a mechanism for the reaction of acetic acid with phosphorus pentachloride.

A carboxylic acid can be converted to an acid chloride by replacing the hydroxyl portion with a chlorine atom. This can be accomplished by using any one of three reagents: PCl_5, PCl_3, or $SOCl_2$. The general reaction can be written as:

$$\underset{\text{Carboxylic acid}}{R-\overset{O}{\overset{\|}{C}}OH} \xrightarrow[\text{or } SOCl_2\text{/ether}]{PCl_5 \text{ or } PCl_3} \underset{\text{Acid chloride}}{R-\overset{O}{\overset{\|}{C}}-Cl}$$

The reaction of acetic acid with phosphorus pentachloride (PCl_5) proceeds through the following mechanisms:

$$CH_3-\overset{O}{\overset{\|}{C}}-\ddot{O}-H + \underset{Cl}{PCl_4} \rightarrow CH_3-\overset{O}{\overset{\|}{C}}-O-PCl_4 + H^+Cl^-$$

$$CH_3-\overset{O}{\overset{\|}{C}}-O-\underset{Cl}{PCl_3} \rightarrow CH_3\overset{O}{\overset{\|}{C}}+ + Cl^- + POCl_3$$

$$CH_3\overset{O}{\overset{\|}{C}}+ + Cl^- \rightarrow CH_3\overset{O}{\overset{\|}{C}}-Cl$$

The mechanism accounts for the production of an acylium ion ($R-C=O$), which reacts with a chloride anion to give the acid chloride. +

Describe the nucleophilic substitution reaction of acyl chlorides.

Acyl chlorides are readily attacked by nucleophiles at the carbonyl carbon. The resulting adduct eliminates the chlorine ion, so the overall reaction is a nucleophilic substitution.

$$R\text{-}\underset{\underset{O}{\|}}{C}\text{-}Cl + :N \longrightarrow R\text{-}\overset{\overset{N}{|}}{\underset{\underset{O^-}{|}}{C}}\text{-}Cl \longrightarrow R\text{-}\underset{\underset{O}{\|}}{C}\text{-}N + Cl^-$$

Nucleophilic attack	Elimination of chloride	Substitution product

Water, alcohols, ammonia, or amines are the nucleophiles that are usually employed in reactions with acyl chlorides. The products of these reactions are carboxylic acids, esters, or carboxamides according to the following reactions.

$$R\text{-}\underset{\underset{O}{\|}}{C}\text{-}Cl + H_2O \rightarrow R\text{-}COOH + HCl$$

Carboxylic acid

$$R\text{-}\underset{\underset{O}{\|}}{C}\text{-}Cl + R'\text{-}OH \rightarrow R\text{-}\underset{\underset{O}{\|}}{C}\text{-}OR' + HCl$$

Alcohol → Ester

$$R\text{-}\underset{\underset{O}{\|}}{C}\text{-}Cl + 2NH_3 \rightarrow R\text{-}\underset{\underset{O}{\|}}{C}\text{-}NH_2 + \overset{+}{NH_4}\overset{-}{Cl}$$

Carboxamide

$$R\text{-}\underset{\underset{O}{\|}}{C}\text{-}Cl + 2R'\text{-}NH_2 \rightarrow R\text{-}\underset{\underset{O}{\|}}{C}\text{-}NH\text{-}R' + R'NH_3^+Cl^-$$

Primary amine → N-Substituted carboxamide

$$R\text{-}\underset{\underset{O}{\|}}{C}\text{-}Cl + 2R'\text{-}NH\text{-}R'' \rightarrow R\text{-}\underset{\underset{O}{\|}}{C}\text{-}N\begin{matrix} R' \\ R'' \end{matrix} + \overset{+}{R'NH_2R''}\ \overset{-}{Cl}$$

Secondary amine → N,N-Disubstituted carboxamide

The mechanism of all these reactions follows the general scheme of nucleophilic substitution. As an example, the reaction of an acyl chloride with an alcohol is shown:

(1) The nucleophile attacks the acyl chloride:

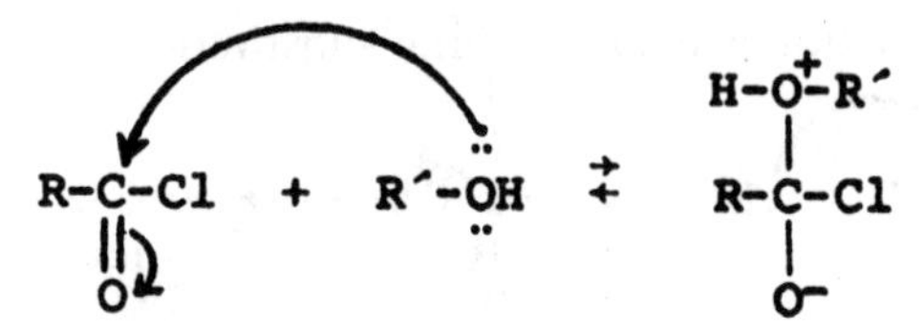

(2) The chloride ion is eliminated:

$$R{-}C(O^-)(Cl){-}\overset{+}{O}(H){-}R' \rightleftharpoons R{-}C(=O){-}\overset{+}{O}(H){-}R' + Cl^-$$

(3) A hydrogen ion is lost to give the ester product:

$$R{-}C(=O){-}\overset{+}{O}(H){-}R' \rightleftharpoons R{-}C(=O){-}O{-}R' + H^+$$

An ester

Although each step of the reaction is reversible, the reaction by-product is hydrogen chloride gas, which usually escapes from the reaction vessel. Thus, the equilibrium reaction is driven completely in a forward direction.

13.2 Carboxylic Acid Anhydrides

Nomenclature (IUPAC System)

When naming acid anhydrides, the word "acid" in the carboxylic acid is replaced by the word "anhydride."

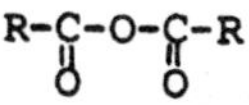

$R-C(=O)-O-C(=O)-R$
Acid anhydride

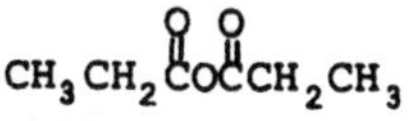

$CH_3CH_2COCCH_2CH_3$ (with C=O on both carbonyl carbons)
Propionic anhydride

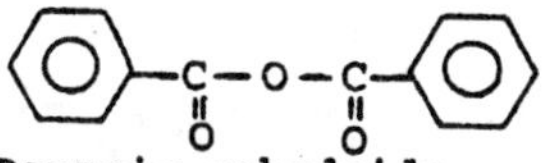

$C_6H_5-C(=O)-O-C(=O)-C_6H_5$
Benzoic anhydride

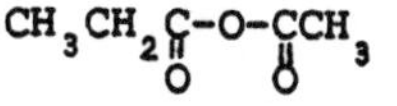

$CH_3CH_2C(=O)-O-C(=O)CH_3$
Acetic propanoic anhydride
(a mixed anhydride)

$(ClCH_2CH_2CH_2CO)_2O$
4-Chlorobutanoic anhydride

Preparation of Acid Anhydrides

A) Carboxylic acid anhydrides are derived from carboxylic acids by the loss of water.

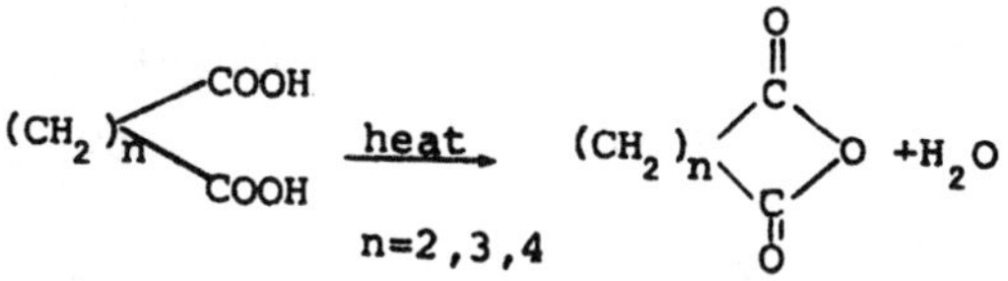

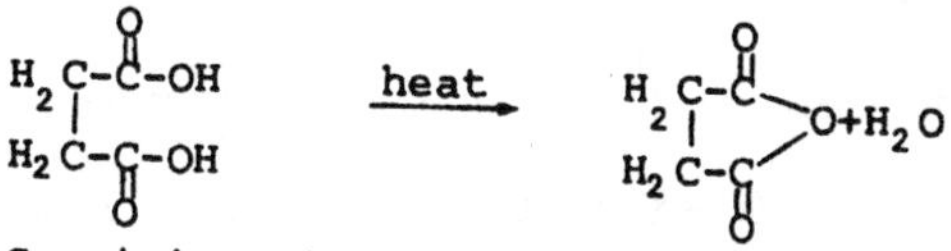

$$(CH_2)_n(COOH)_2 \xrightarrow{heat} (CH_2)_n(C{=}O)_2O + H_2O \quad n=2,3,4$$

$$H_2C(-C(=O)-OH)-H_2C(-C(=O)-OH) \xrightarrow{heat} \text{succinic anhydride} + H_2O$$

Succinic acid → Succinic anhydride

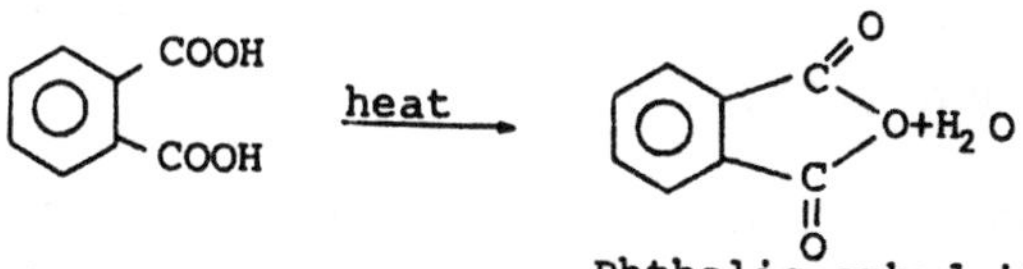

$$C_6H_4(COOH)_2 \xrightarrow{heat} C_6H_4(C{=}O)_2O + H_2O$$

Phthalic acid → Phthalic anhydride

B) Acid chlorides react with salts of organic acids to yield anhydrides.

$$R-\underset{\underset{O}{\|}}{C}-Cl + R'-\underset{\underset{O}{\|}}{C}-O^-Na^+ \rightarrow R-\underset{\underset{O}{\|}}{C}-O-\underset{\underset{O}{\|}}{C}-R' + NaCl$$

acid chloride — carboxylic acid salt — acid anhydride

Example

$$CH_3COCl + CH_3COONa \rightarrow (CH_3CO)_2O + NaCl$$

acetyl chloride — sodium acetate — acetic anhydride

C) Ketene, $CH_2 = C = O$, reacts with acid to yield anhydrides.

Example

$$CH_3COCH_3 \xrightarrow{700°-750°} CH_2 = C = O + CH_4$$

acetone — ketene

$$CH_2{=}C{=}O + CH_3CH_2COOH \rightarrow CH_3-\overset{\overset{O}{\|}}{C}-O-\overset{\overset{O}{\|}}{C}-CH_2CH_3$$

propanoic acid — acetic propionic anhydride

$$CH_3COOH \xrightarrow[700°]{AlPO_4} H_2O + CH_2 = C = O$$

acetic acid — ketene

$$CH_2 = C = O + CH_3COOH \rightarrow (CH_3CO)_2O$$

acetic acid — acetic anhydride

D) Thallium salts of carboxylic acids react with $SOCl_2$ to give symmetric anhydrides.

$$2RCOOTl + SOCl_2 \rightarrow [(RCOO)_2SO] + 2TlCl \rightarrow (RCO)_2O + SO_2$$

thallium (I) salt — acid anhydride

Reactions of Acid Anhydrides

A) Conversion to acids and acid anhydrides

$$(RCO)_2O + HZ \rightarrow RCOZ + RCOOH$$

a) Conversion into acids. Hydrolysis.

$$CH_3\underset{\underset{O}{\|}}{C}-O-\underset{\underset{O}{\|}}{C}CH_3 + H_2O \longrightarrow 2CH_3COOH$$

Acetic anhydride — Acetic acid

$$C_6H_5-\underset{\underset{O}{\|}}{C}-O-\underset{\underset{O}{\|}}{C}-C_6H_5 + H_2O \longrightarrow 2\,C_6H_5-C(=O)OH$$

Benzoic anhydride — Benzoic acid

b) Conversion into amides. Ammonolysis.

Example

$$(CH_3CO)_2O + 2NH_3 \rightarrow CH_3\underset{\underset{O}{\|}}{C}-NH_2 + CH_3COO^-\overset{+}{N}H_4$$

acetic anhydride — acetamide — ammonium acetate

$$\text{succinic anhydride } (H_2C-C(=O)-O-C(=O)-CH_2) + 2NH_3 \rightarrow \begin{array}{l}CH_2CONH_2\\ |\\ CH_2COO^-NH_4^+\end{array} \xrightarrow{H^+} \begin{array}{l}CH_2CONH_2\\ |\\ CH_2COOH\end{array}$$

succinic anhydride — 1-amino-ammonium succinate — 1-amino-succinic acid (half amide, half acid)

$$(RCO)_2O + R'NH_2 \longrightarrow R\underset{\underset{O}{\|}}{C}-NHR' + RCOOH$$

Acid anhydride — Primary amine — N-Substituted Carboxamide — Carboxylic acid

$$C_6H_5-\underset{\underset{O}{\|}}{C}-O-\underset{\underset{O}{\|}}{C}-C_6H_5 + C_6H_5NH_2 \longrightarrow C_6H_5-\underset{\underset{O}{\|}}{C}-\underset{\underset{H}{|}}{N}-C_6H_5 + C_6H_5-C(=O)OH$$

Benzoic anhydride — Aniline — N-Phenyl benzamide — Benzoic acid

c) Conversion into esters. Alcoholysis.

$$(CH_3CO)_2O + CH_3OH \longrightarrow CH_3COOCH_3 + CH_3COOH$$

Acetic anhydride | Methyl alcohol | Methyl acetate (An ester) | Acetic acid

$$C_6H_5-\underset{O}{\overset{\|}{C}}-O-\underset{O}{\overset{\|}{C}}-C_6H_5 + CH_3CH_2OH \longrightarrow C_6H_5-C(=O)OCH_2CH_3 + C_6H_5-C(=O)OH$$

Benzoic anhydride | Ethyl alcohol | Ethyl benzoate | Benzoic acid

$$\text{Phthalic anhydride} + CH_3CH_2\underset{OH}{CH}CH_3 \longrightarrow C_6H_4(COOCH(CH_3)CH_2CH_3)(COOH)$$

Phthalic anhydride | sec-Butyl alcohol | Sec-Butyl hydrogen phthalate (half ester,half acid)

B) Reduction to alcohols.

Example

$$(CH_3CO)_2O + LiAlH_4 \rightarrow LiAlO(OCH_2CH_3)_2 \xrightarrow{H^+} 2CH_3CH_2OH$$

acetic anhydride | ethyl alcohol

C) Formation of ketones. Friedel-Crafts acylation.

$$(RCO)_2O + ArH \xrightarrow[\text{or other Lewis acid}]{AlCl_3} R-\underset{O}{\overset{\|}{C}}-Ar + RCOOH$$

A ketone

$$(CH_3CO)_2O + C_6H_3(CH_3)_3 \xrightarrow{AlCl_3} CH_3-\underset{O}{\overset{\|}{C}}-C_6H_2(CH_3)_3 + CH_3COOH$$

Acetic anhydride | Mesitylene | Methyl mesityl ketone | Acetic acid

Problem Solving Examples:

Write equations to show the reaction of benzoic anhydride with (a) water, (b) ethyl alcohol, and (c) aniline.

Carboxylic acid anhydrides such as benzoic anhydride undergo the same type of nucleophilic attack and substitution as acyl halides. It may be represented by

$$\text{R-}\underset{\substack{\|\\ \text{O}}}{\text{C}}\text{-X} + \text{:N (nucleophile)} \rightarrow \text{R-}\overset{\substack{\text{N}\\ |}}{\underset{\substack{|\\ \text{O}^-}}{\text{C}}}\text{-X} \quad \text{R-}\underset{\substack{\|\\ \text{O}}}{\text{C}}\text{-N} + \text{X}^-$$

where X may be Cl, NH_2, NHR, NR_2, $\underset{\substack{\|\\ \text{O}}}{\text{OC}}\text{-R}$, or OR.

The attack of the nucleophile, which is usually water, ammonia or amines, and alcohols, causes the leaving of X and then gives the substituted compound.

Hence, the following equations for the reaction of benzoic anhydride

$$\left[C_6H_5\text{-}\underset{\substack{\|\\ \text{O}}}{\text{C}}\text{-O-}\underset{\substack{\|\\ \text{O}}}{\text{C}}\text{-}C_6H_5 \right]$$

with the various nucleophiles given can be written:

(a)

$$C_6H_5\text{-}\underset{\substack{\|\\ \text{O}}}{\text{C}}\text{-O-}\underset{\substack{\|\\ \text{O}}}{\text{C}}\text{-}C_6H_5 + H_2O \text{ (water)} \rightarrow \underbrace{C_6H_5\text{-}\underset{\substack{\|\\ \text{O}}}{\text{C}}\text{-OH} + C_6H_5\text{-}\underset{\substack{\|\\ \text{O}}}{\text{C}}\text{-OH}}_{\text{Benzoic acid}}$$

(b)

$$C_6H_5-\overset{O}{\overset{||}{C}}-O-\overset{O}{\overset{||}{C}}-C_6H_5 + CH_3CH_2OH \text{ (ethyl alcohol)} \longrightarrow$$

$$C_6H_5-\overset{O}{\overset{||}{C}}-OCH_2CH_3 + C_6H_5-\overset{O}{\overset{||}{C}}-OH$$

Ethyl benzoate   Benzoic acid

(c)

$$C_6H_5-\overset{O}{\overset{||}{C}}-O-\overset{O}{\overset{||}{C}}-C_6H_5 + C_6H_5NH_2 \text{ (aniline)} \longrightarrow$$

$$C_6H_5-\overset{O}{\overset{||}{C}}-\overset{H}{N}-C_6H_5 + C_6H_5-\overset{O}{\overset{||}{C}}-OH$$

N-phenyl benzamide   Benzoic acid

Give structural formulas for compounds A through G.

Benzene + succinic anhydride

$$\text{Benzene} + \underset{\text{Succinic Acid}}{\text{COOH–CH}_2\text{–CH}_2\text{–COOH}} \xrightarrow{AlCl_3} \xrightarrow{H_2O} A(C_{10}H_{10}O_3)$$

$$A + Zn(Hg) \xrightarrow{HCl} B(C_{10}H_{12}O_2)$$

$$B + SOCl_2 \longrightarrow C(C_{10}H_{11}OCl)$$

$$C \xrightarrow{AlCl_3} D(C_{10}H_{10}O)$$

$$D + H_2 \xrightarrow{Pt} E(C_{10}H_{12}O)$$

$$E + H_2SO_4 \xrightarrow{heat} F(C_{10}H_{10})$$

$$F \xrightarrow{Pt,\ heat} G(C_{10}H_8) + H_2$$

A The sequence of reactions depicted in this problem represent the Haworth synthesis of naphthalene. It commences with the addition of $AlCl_3$ to a mixture of benzene and succinic anhydride to produce $A(C_{10}H_{10}O_3)$. This reaction is an example of a Friedel-Crafts acylation. This is a modification of the Friedel-Crafts alkylation, which is the most important method for attaching alkyl side chains to aromatic rings. In Friedel-Crafts acylation, an acyl group (RCO—) is attached to the aromatic ring; this requires the generation of an electrophile, the acylium ion (R – C = O). The acylium ion will undergo an electrophilic aromatic substitution reaction with the aromatic ring to form an aromatic ketone. The acylium ion is formed by using a Lewis acid (i.e., $AlCl_3$) as a catalyst. Acid anhydrides will react with $AlCl_3$ to give an acylium ion. The general mechanism of an acid anhydride reacting with an aromatic ring in the presence of $AlCl_3$ is:

$$\text{RC(=O)OC(=O)R}' + \text{AlCl}_3 \rightarrow \text{RC}^{+}\text{(=O)} + \text{Cl}_3\bar{\text{Al}}\text{OC(=O)R}'$$

acid anhydride

acylium ion

$$\text{RC}^{+}\text{(=O)} + \text{C}_6\text{H}_6 \rightarrow \text{RC(=O)C}_6\text{H}_5$$

benzene (aromatic ring)

aromatic ketone

$$\text{succinic anhydride: } CH_2\text{-C(=O)-O-C(=O)-}CH_2 \text{ (cyclic)}$$

Succinic anhydride, is an acid anhydride and should react with benzene and $AlCl_3$ as depicted above. But note that succinic anhydride is a cyclic acid anhydride; reaction with $AlCl_3$ will open the ring to give the acylium ion and the $-OACl_3$ portion on the same molecule. This will react with benzene to give an aromatic ketone. Subsequent quenching of the product with water will convert the $-OACl_3$ to an $-OH$. The mechanism is:

CH2-C(=O)-:O:-C(=O)-CH2 (cyclic) + AlCl3 → CH2-C(=O)+ / CH2-C(=O)-OAlCl3 (−)

↓ benzene

HOCCH2CH2C(=O)-C6H5 ← (H2O, − AlCl3) ← CH2-C(=O)-C6H5 / CH2-C(=O)-OAlCl3 (−)

β-Benzoylpropionic acid
(Product A)

Product A now reacts with Zn(Hg) in the presence of HCl to give B($C_{10}H_{12}O_2$). This reaction is a Clemmensen reduction; it converts an acyl group into an alkyl group. Therefore,

β-Benzoylpropionic acid (Product A) —Zn(Hg), HCl→ γ-phenylbutyric acid (Product B)

β-Benzoylpropionic acid
(Product A)

γ-phenylbutyric acid
(Product B)

Product B is then reacted with $SOCl_2$ to generate product C ($C_{10}H_{11}OCl$). Thionyl chloride ($SOCl_2$) is a reagent used to convert carboxylic acids into acid chlorides as can be seen:

$$\text{HOC(=O)CH}_2\text{CH}_2\text{CH}_2\text{-C}_6\text{H}_5 \xrightarrow{SOCl_2} \text{ClC(=O)CH}_2\text{CH}_2\text{CH}_2\text{-C}_6\text{H}_5$$

γ-phenylbutyric acid (Product B) → γ-phenylbutyl chloride (Product C)

Product C is transformed into product D ($C_{10}H_{10}O$) by $AlCl_3$. This suggests another Friedel-Crafts acylation. However, this time a ring closure results as illustrated:

$$\text{ClC(=O)CH}_2\text{CH}_2\text{CH}_2\text{-C}_6\text{H}_5 \xrightarrow{AlCl_3} \text{tetralone} + HCl$$

γ-phenyl butyl chloride (Product C) → γ-tetralone (Product D)

The addition of hydrogen in the presence of Pt to product D indicates product E ($C_{10}H_{12}O$) results from catalytic hydrogenation of the carbonyl group to a hydroxyl group.

$$\text{tetralone} + H_2 \xrightarrow{Pt} \text{tetralol (OH)}$$

α-tetralone (Product D) → α-tetralol (Product E)

Alcohols in the presence of a mineral acid such as H_2SO_4 can be dehydrated to alkenes. This is exactly what occurs in the generation of product F ($C_{10}H_{10}$) as indicated:

$$\text{tetralone} + H_2 \xrightarrow{Pt} \text{tetralol (OH)}$$

α-tetralone (Product D) → α-tetralol (Product E)

The final reaction in this sequence is aromatization. The addition of Pt and heat results in a dehydrogenation that produces product G, naphthalene.

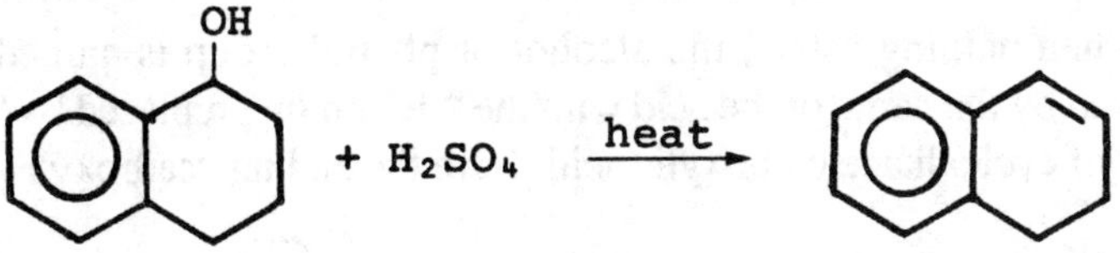

OH

+ H_2SO_4 $\xrightarrow{\text{heat}}$

α-tetralol
(Product E)

α-tetralene
(Product F)

Note: a-tetralone

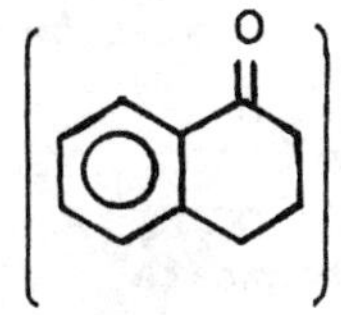

O

could have been converted to tetralin

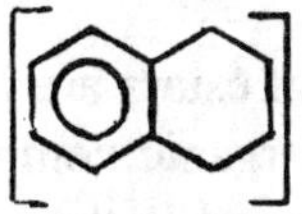

via a Clemmensen reduction. The tetralin could then undergo aromatization to yield naphthalene.

13.3 Esters

Nomenclature (IUPAC System)

When naming esters, the alcohol or phenol group is named first, followed by the name of the acid with the "-ic" ending replaced by "-ate." Esters of cycloalkane carboxylic acids have the ending "carboxylate."

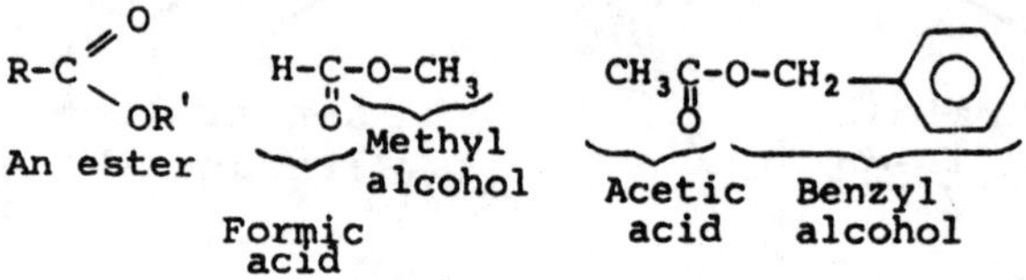

$R-C(=O)-OR'$
An ester

$H-C(=O)-O-CH_3$
Formic acid, Methyl alcohol
Methyl formate

$CH_3C(=O)-O-CH_2-C_6H_5$
Acetic acid, Benzyl alcohol
Benzyl acetate

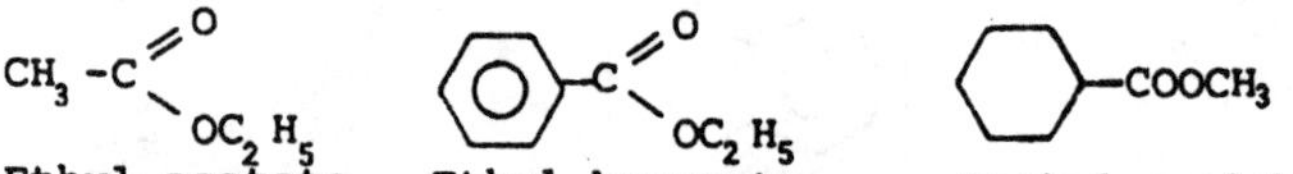

$CH_3-C(=O)-OC_2H_5$
Ethyl acetate

$C_6H_5-C(=O)-OC_2H_5$
Ethyl benzoate

$C_6H_{11}-COOCH_3$
Methyl cyclohexane-carboxylate

Properties

The low molecular weight esters are insoluble in water. They are excellent solvents for many organic compounds. Volatile esters have pleasant odors, and they are used in the preparation of perfumes and artificial flavorings.

Preparation of Esters

A) From acids

$$RCOOH + R'OH \underset{}{\overset{H^+}{\rightleftharpoons}} RCOOR' + H_2O$$

Carboxylic acid (R may be alkyl or aryl); Alcohol (R' is usually alkyl); Ester

Reactivity of R'OH

1° > 2° > 3°

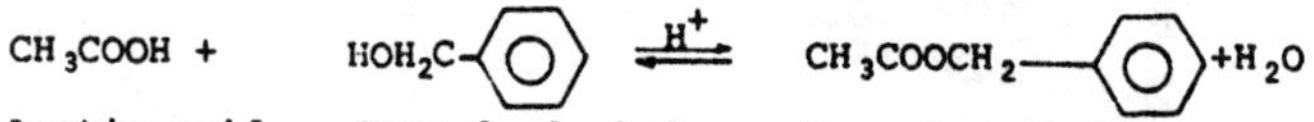

$$CH_3COOH + HOH_2C-C_6H_5 \overset{H^+}{\rightleftharpoons} CH_3COOCH_2-C_6H_5 + H_2O$$

Acetic acid; Benzyl alcohol; Benzyl acetate

$$C_6H_5-COOH + HOCH_2CH(CH_3)CH_3 \overset{H^+}{\rightleftharpoons} C_6H_5-COOCH_2CH(CH_3)CH_3 + H_2O$$

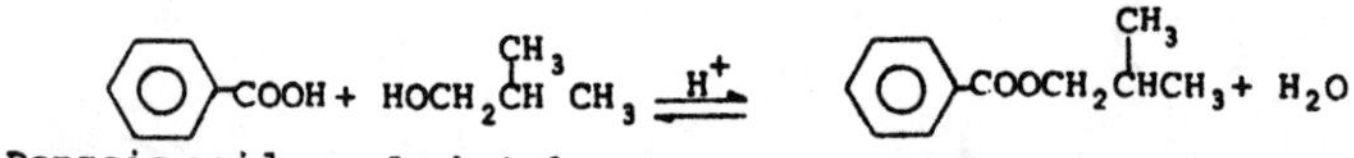

Benzoic acid; Isobutyl alcohol; Isobutyl benzoate

B) From acid chlorides

$$R-\underset{\underset{O}{\|}}{C}-Cl + R'OH \text{ (or ArOH)} \longrightarrow R-\underset{\underset{O}{\|}}{C}-O-R' \text{ (or RCOOAr)} + HCl$$

$$\text{o-Bromobenzoyl chloride } (o\text{-}BrC_6H_4COCl) + C_2H_5OH \xrightarrow{\text{Pyridine}} \text{Ethyl o-bromobenzoate } (o\text{-}BrC_6H_4COOC_2H_5) + HCl$$

C) From anhydrides.

$$(RCO)_2O + R'OH \text{ (or ArOH)} \longrightarrow RCOOR' \text{ (or RCOOAr)} + RCOOH$$

$$CH_3\underset{\underset{O}{\|}}{C}-O-\underset{\underset{O}{\|}}{C}CH_3 + CH_3-OH \longrightarrow CH_3\underset{\underset{O}{\|}}{C}-OCH_3 + CH_3COOH$$

Acetic anhydride — Methyl alcohol — Methyl acetate

$$(CH_3CO)_2O + HO-C_6H_4-NO_2 \xrightarrow{\text{NaOH}} CH_3COO-C_6H_4-NO_2 + CH_3COOH$$

Acetic anhydride — p-Nitrophenol — p-Nitrophenyl acetate

D) From salts of very active or heavy metals and alkyl halides.

$$R-CO\overset{-}{O}\overset{+}{M} + X-R' \text{ Æ } R-CO-O-R' + MX$$

Example

$$CH_3-CO-\overset{-}{O}\overset{+}{A}g + I-CH_2CH_3 \rightarrow CH_3-CO-O-CH_2CH_3 + AgI$$

ethyl acetate

E) From esters. Transesterification.

$$RCOOR' + R''OH \underset{}{\overset{H^+ \text{ or } OR^-}{\rightleftharpoons}} RCOOR'' + R'OH$$

F) From alcohols and ketene.

$$CH_2=C=O + ROH \rightarrow CH_3COOR$$

ketene — ester

G) Methyl esters from diazomethane.

$$RCOOH + CH_2N_2 \rightarrow \underset{\text{methyl ester}}{RCOOCH_3} + N_2$$

H) Preparation of lactones (cyclic esters) and lactides (dilactones). Hydroxy acids undergo self-esterification under acid catalysis.

$$\underset{\gamma\text{-Hydroxybutyric acid}}{HOCH_2CH_2CH_2COOH} \xrightleftharpoons{H^+} \underset{\gamma\text{-Butyrolactone}}{\text{(cyclic lactone)}} + H_2O$$

$$\underset{\text{Lactic acid}}{2CH_3CH(OH)COOH} \xrightarrow[-H_2O]{H^+} CH_3CH(OH)COOCH(CH_3)COOH \xrightarrow{-H_2O} \underset{\text{Lactide}}{\text{(cyclic dilactone, } H_3C, CH_3)}$$

Reactions of Esters

A) Conversion into acids and acid derivatives.

a) Acid-catalyzed hydrolysis of esters.

$$R-\underset{\underset{O}{\|}}{C}-OR' + H_2O \text{ (excess)} \xrightarrow{H^+} R-\underset{\underset{O}{\|}}{C}-OH + R'-OH$$

$$\underset{\text{Ethyl benzoate}}{C_6H_5COOC_2H_5} + H_2O \xrightarrow{H_2SO_4} \underset{\text{Benzoic acid}}{C_6H_5COOH} + \underset{\text{Ethyl alcohol}}{C_2H_5OH}$$

b) Basic hydrolysis. Saponification of esters.

$$R-\underset{\underset{O}{\|}}{C}-OR' + H_2O \xrightarrow{OH^-} R-\underset{\underset{O}{\|}}{C}-O^- + R'-OH$$

$$\xrightarrow{H^+} RCOOH$$

$$\underset{\text{Ethyl benzoate}}{C_6H_5COOC_2H_5} + H_2O \xrightarrow{NaOH} \underset{\text{Sodium benzoate}}{C_6H_5COO^-Na^+} + \underset{\text{Ethyl alcohol}}{C_2H_5OH}$$

$$CH_3CH_2COOCH_3 + H_2O \xrightarrow{NaOH} CH_3CH_2COO^-Na^+ + CH_3OH$$

c) Conversion into amides. Ammonolysis.

$$R-\underset{\underset{O}{\|}}{C}-OR' + NH_3 \rightarrow R-\underset{\underset{O}{\|}}{C}-NH_2 + R'-OH$$

Example

$$\underset{\text{ethyl acetate}}{CH_3COOC_2H_5} + NH_3 \rightarrow \underset{\text{acetamide}}{CH_3CONH_2} + \underset{\text{ethyl alcohol}}{C_2H_5OH}$$

d) Conversion into esters. Transesterification. Alcoholysis.

$$R-\underset{\underset{O}{\|}}{C}-OR' + R''-OH \xrightleftharpoons[]{\text{acid or base}} R-\underset{\underset{O}{\|}}{C}-OR'' + R'-OH$$

Example

$$\underset{\text{a glyceride (a fat)}}{\begin{array}{l} CH_2-O-\underset{\underset{O}{\|}}{C}-R \\ | \\ CH-O-\underset{\underset{O}{\|}}{C}-R' \\ | \\ CH_2-O-\underset{\underset{O}{\|}}{C}-R'' \end{array}} + CH_3OH \xrightarrow{\text{acid or base}} \underset{\text{Mixture of methyl esters}}{\begin{array}{c} RCOOCH_3 \\ + \\ R'COOCH_3 \\ + \\ R''COOCH_3 \end{array}} + \underset{\text{glycerol}}{\begin{array}{l} CH_2OH \\ | \\ CHOH \\ | \\ CH_2OH \end{array}}$$

B) Reaction with Grignard reagents.

$$RCOOR' + 2R''MgX \rightarrow R-\underset{OH}{\overset{R''}{\underset{|}{\overset{|}{C}}}}-R''$$

Example — tertiary alcohol

$$CH_3-\overset{CH_3}{\overset{|}{C}}HCOOC_2H_5 + 2CH_3MgI \rightarrow CH_3\overset{CH_3}{\overset{|}{C}}H-\underset{OH}{\underset{|}{\overset{CH_3}{\overset{|}{C}}}}-CH_3$$

ethyl isobutyrate — methyl magnesium iodide 2 moles — 2,3-dimethyl-2-butanol

C) Reduction to alcohols.

a) Catalytic hydrogenation. Hydrogenolysis.

$$RCOOR' + 2H_2 \xrightarrow[250°C]{CuO, CuCr_2O_4} RCH_2OH + R'OH$$

3000-6000 lb /in^2 — primary alcohol

Example

$$CH_3-\underset{CH_3}{\underset{|}{\overset{CH_3}{\overset{|}{C}}}}-COOC_2H_5 + 2H_2 \xrightarrow[250°, 3300\ lb/in^2]{CuO, CuCr_2O_4}$$

ethyl trimethyl acetate (ethyl 2,2-dimethyl propanoate)

$$CH_3-\underset{CH_3}{\underset{|}{\overset{CH_3}{\overset{|}{C}}}}-CH_2OH + C_2H_5OH$$

Neopentyl alcohol (2,2-dimethyl propanol) — ethyl alcohol

b) Bouvaeult-Blance method.

$$RCOOR' \xrightarrow{\text{Na + alcohol}} RCH_2OH + R'OH$$

c) Lithium aluminum hydride reduction.

$$4RCOOR' + 2LiAlH_4 \xrightarrow[\text{ether}]{\text{anhyd}} LiAl(OCH_2R)_4 + LiAl(OR')_4 \xrightarrow{H^+} RCH_2OH + R'OH$$

Example

$$CH_3(CH_2)_7CH = CH(CH_2)_7COOCH_3 \xrightarrow{LiAlH_4} CH_3(CH_2)_7CH = CH(CH_2)_7CH_2OH$$

methyl oleate (methyl cis-9-octadecenoate) → oleyl alcohol (cis-9-octadecene-1-ol)

D) Reaction with carbanions. Claisen condensation.

$$-C(=O)OR' + -\overset{|}{\underset{H}{C}}-C(=O)OR' \xrightarrow{^-OC_2H_5} -C(=O)-\overset{|}{\underset{|}{C}}-C(=O)OR'$$

A β-keto ester

$$C_6H_5COOC_2H_5 + CH_3COOC_2H_5 \xrightarrow{^-OC_2H_5} C_6H_5C(=O)-CH_2COOC_2H_5 + C_2H_5OH$$

Ethyl benzoate + Ethyl acetate → Ethyl benzoyl acetate + C_2H_5OH

$$C_2H_5O-C(=O)-OC_2H_5 + C_6H_5CH_2COOC_2H_5 \xrightarrow{C_2H_5O^-} C_2H_5O-C(=O)-CH(C_6H_5)COOC_2H_5 + C_2H_5OH$$

Ethyl carbonate + Ethyl phenylacetate → Ethyl phenylmalonate (Phenylmalonic ester)

$$2CH_3C(=O)OCH_2CH_3 \xrightarrow{C_2H_5O^-} CH_3C(=O)CH_2COOCH_2CH_3$$

Ethyl acetate (2 moles) → Ethyl acetoacetate (Aceto acetic ester)

Problem Solving Examples:

Show how the following alcohols react with carboxylic acids to yield esters.

(a) CH_3CH_2OH (b) C_6H_5–CH_2OH

Ethanol Benzyl alcohol

Alcohols react with carboxylic acids in the presence of mineral acid catalysts, to give esters. The general reaction is:

$$R\text{-}OH + R'\text{-}\overset{O}{\overset{\|}{C}}OH \underset{}{\overset{H_2SO_4}{\rightleftharpoons}} R'\text{-}\overset{O}{\overset{\|}{C}}\text{-}O\text{-}R + H_2O$$

where the mineral acid is almost always sulfuric acid, H_2SO_4. Also, this type of reaction is completely reversible, in that the same catalyst, hydrogen ion, that catalyzes the forward reaction, esterification, necessarily catalyzes the reverse reaction, hydrolysis. The mechanism by which the forward reaction works is direct esterification where molecules of water are "split out" when the alcohol and carboxylic acid are mixed together. For example:

$$RCO\boxed{OH} + \boxed{H}OR' \rightarrow RCOOR' + H_2O$$

With this principle, the general explanation, and reaction, carboxylic acids can be added to each alcohol to yield an ester.

(a)

$$CH_3CH_2OH + C_6H_5\text{-}CH_2CH_2CH_2\overset{O}{\overset{\|}{C}}\text{-}OH \underset{}{\overset{H_2SO_4 \text{ reflux}}{\rightleftharpoons}}$$

Ethanol γ-Phenylbutyric acid

$$C_6H_5\text{-}CH_2CH_2CH_2\overset{O}{\overset{\|}{C}}\text{-}OCH_2CH_3 + H_2O$$

Ethyl-γ-Phenylbutyrate

(b)

$$C_6H_5\text{-}CH_2OH + CH_3COOH \xrightarrow{\text{conc. } H_2SO_4} CH_3COOCH_2\text{-}C_6H_5 + H_2O$$

Benzyl Alcohol — Acetic acid — Benzyl acetate

What product do you expect from the following reaction?

$$\text{Phthalic anhydride} + \underset{(1\text{ mole})}{CH_3OH} \rightarrow$$

The yield of the product is essentially quantitative. Write a mechanism for the reaction.

When an acid anhydride is treated with an alcohol, an esterification reaction occurs. The general reaction is:

$$RC(=O)\text{-}O\text{-}C(=O)R + R'OH \rightarrow RC(=O)OH + R'OC(=O)R$$

Esterification of this type involves the breaking of the acyl-oxygen bond:

$$R\text{-}C(=O)\text{-}O\text{-}C(=O)\text{-}R$$

(This is the carbon being attacked)

acyl-oxygen bond (broken)

The reaction of a cyclic acid anhydride with an alcohol produces a molecule with both an ester and carboxylic acid functionality. The mechanism for the reaction of phthalic anhydride with methanol is:

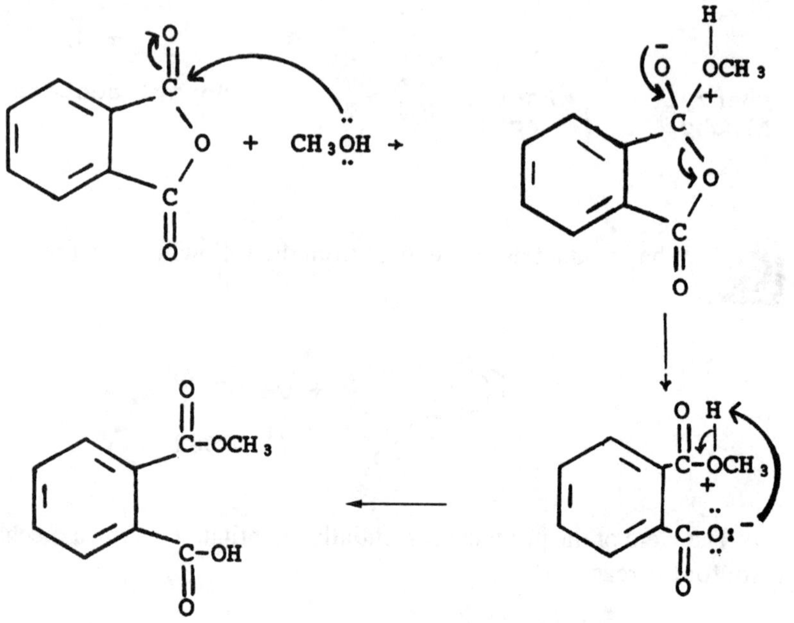

O + CH₃OH →

H
O OCH₃
C +
O
C
O

O H
C–OCH₃
+
C–O:⁻
O

O
C–OCH₃
C–OH
O

Only one mole of methanol is used in this reaction. If two or more moles had been used in the presence of acid, the product would have the methyl diester. The reaction would be:

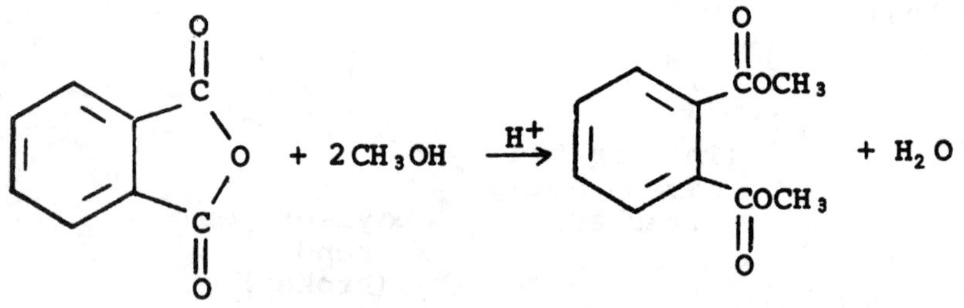

$$\text{phthalic anhydride} + 2\,CH_3OH \xrightarrow{H^+} C_6H_4(COOCH_3)_2 + H_2O$$

13.4 Amides

Nomenclature (IUPAC System)

When naming amides, the "-ic acid" of the common name (or the "-oic acid" of the IUPAC name) of the parent acid is replaced by "amide." Amides of cycloalkane carboxylic acids have the ending carboxamide.

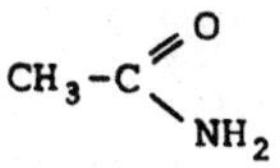

$CH_3-C(=O)-NH_2$
Acetamide
Ethanamide

$H_2NOCCH_2CH_2CONH_2$
1,4-Butane diamide

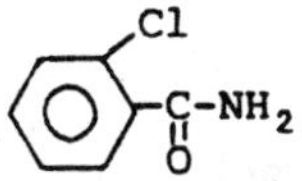

$C_6H_4(Cl)-C(=O)-NH_2$
o-Chlorobenzamide

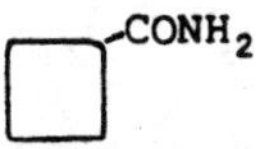

$C_4H_7-CONH_2$
Cyclobutane-carboxamide

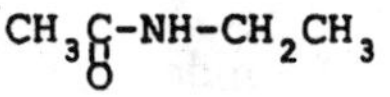

$CH_3C(=O)-NH-CH_2CH_3$
N-Ethyl acetamide

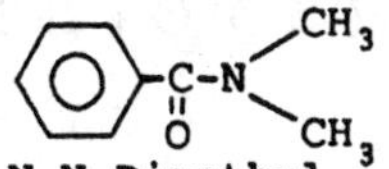

$C_6H_5-C(=O)-N(CH_3)_2$
N,N-Dimethyl benzamide

Properties of Amides

Amides have high boiling points due to strong intermolecular hydrogen bonding. Amides with up to five or six carbons are soluble in water. Like the other acid derivatives, they are soluble in the usual organic solvents. Primary amides are very weak basic compounds and are insoluble in dilute acids.

Preparation of Amides

A) Amides from acid chlorides and ammonia or amines.

$$RCOCl + 2NH_3 \rightarrow RCONH_2 + NH_4Cl$$

amide (1°)

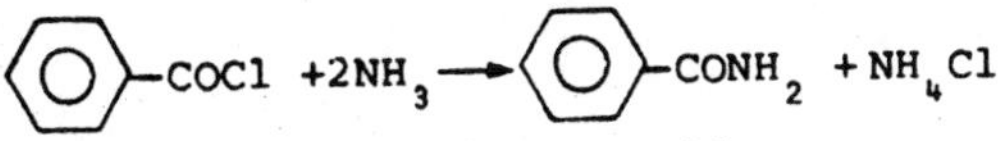

$$C_6H_5-COCl + 2NH_3 \rightarrow C_6H_5-CONH_2 + NH_4Cl$$

Benzoyl chloride → Benzamide

$$RCOCl + R'NH_2 \rightarrow RCONHR' + HCl$$

1° amine 2° amide

$$RCOCl + 2NHR'_2 \rightarrow RCONR'_2 + R'_2{}^+NH_2Cl^-$$

2° amine 3° amide

B) From acid anhydrides and ammonia or amines.

$$(RCO)_2O + 2NH_3 \rightarrow RCONH_2 + RCOONH_4$$

amide (1°)

Example

$$(CH_3CO)_2O + 2NH_3 \rightarrow CH_3CONH_2 + CH_3COO^-NH_4^+$$

acetic anhydride acetamide ammonium acetate

$$(RCO)_2O + 2R'NH_2 \rightarrow RCONHR' + R'NH_3{}^+RCO_2{}^-$$

1° amine 2° amide

$$(RCO)_2O + NHR'_2 \rightarrow RCONR'_2 + RCOOH$$

2° amine 3° amide

C) Amides from esters by ammonolysis.

$$R{-}\underset{\substack{\|\\ O}}{C}{-}O{-}R' + NH_3 \rightarrow RCONH_2 + R'OH$$

1° amide

D) Amides by the pyrolysis of ammonium carboxylates.

$$R{-}\underset{\substack{\|\\ O}}{C}{-}OH + NH_3 \rightarrow R{-}\underset{\substack{\|\\ O}}{C}{-}O^- \overset{+}{N}H_4 \xrightarrow{heat} R{-}\underset{\substack{\|\\ O}}{C}{-}NH_2 + H_2O$$

ammonium carboxylate 1° amide

E) From nitriles.

$$RC\equiv N + H_2O \xrightarrow[\text{or } OH^-]{H^+} RCONH_2$$

1° amide

Example

$$CH_3CH_2C\equiv N + H_2O \xrightarrow[\text{conc.}]{H_2SO_4} CH_3CH_2CONH_2$$

propanamide

F) Amides from amines and ketene.

$$RNH_2 + CH_2 = C = O \rightarrow CH_3CONHR$$

2° amide

G) Beckmann rearrangement.

$$\underset{\text{R}-\overset{\|}{C}R'}{:N-OH} \xrightarrow{H^+} \underset{R-\overset{\|}{C}-R'}{:N-\overset{+}{O}H_2} \longrightarrow R-N=C^+-R' \xrightarrow{H_2O}$$

R group trans
to OH migrates

$$\underset{\;\;\;\;\;\;\;\;\;{}^{+}OH_2}{R-N=C-R'} \xrightarrow{-H^+} \underset{\;\;\;\;\;\;H-O}{R-N=C-R'} \rightleftharpoons \underset{\;\;\;\;\;H\;\;\;O}{R-N-C-R'}$$

enol form of amide

more stable keto form of amide

H) Preparation of secondary and tertiary amides.

$$RCONH_2 + (CH_3CO)_2O \xrightarrow{\text{heat}} RCONHCOCH_3 \xrightarrow[\text{heat}]{(CH_3CO)_2O}$$

2° amide

$$RCON(COCH_3)_2$$

3° amide

I) Preparation of lactams (cyclic amides).

$$H_2NCH_2CH_2CH_2COOH \xrightarrow{Heat} \text{(cyclic lactam)} + H_2O$$

γ-Butyrolactam

Reactions of Amides

A) Hydrolysis.

$$RCONH_2 + H_2O \xrightarrow{H^+} RCOOH + NH_4^+$$

$$RCONH_2 + H_2O \xrightarrow{OH^-} RCOO^- + NH_3$$

$$C_6H_5CONH_2 + H_2SO_4 + H_2O \longrightarrow C_6H_5COOH + NH_4^+ HSO_4^-$$

Benzamide → Benzoic acid

$$CH_3CH_2CH_2CONH_2 + NaOH + H_2O \longrightarrow CH_3CH_2CH_2COO^-Na^+ + NH_3$$

Butyramide → Sodium butyrate

B) Conversion into imides.

$$\text{Phthalic anhydride} \xrightarrow{NH_3, heat} \text{Phthalimide}$$

Phthalic anhydride Phthalimide

C) Salt formation. Gabriel's synthesis.

Primary amides form salts that are hydrolyzed by the stronger base water. The hydrogen of secondary amides is more acidic and the stable salts are formed with aqueous NaOH.

$$\text{N-Alkylphthalimide} \xrightarrow[NaOH]{H_2O} C_6H_4(COONa)_2 + RNH_2$$

N-Alkylphthalimide Sodium phthalate

D) Reaction with nitrous acid.

$$RCONH_2 + ONOH \rightarrow RCOOH + N_2 + H_2O$$

E) Dehydration of primary amine into nitrile by treatment with a strong dehydrating agent.

$$RCONH_2 + P_2O_5 \rightarrow RC \equiv N + 2HPO_3$$

F) Reduction of amides by lithium aluminum hydride.

$$R-\underset{\underset{O}{\|}}{C}-NH_2 \xrightarrow{LiAlH_4} \xrightarrow{H_2O} RCH_2-NH_2$$

primary amide → primary amine

$$R-\underset{\underset{O}{\|}}{C}-NH-R' \xrightarrow{LiAlH_4} \xrightarrow{H_2O} R-CH_2-NH-R'$$

N-substituted amide
secondary amide → secondary amine

$$R-\underset{\underset{O}{\|}}{C}-NR'_2 \xrightarrow{LiAlH_4} \xrightarrow{H_2O} R-CH_2-N-R'_2$$

N,N-disubstituted
amide
tertiary amide → tertiary amine

G) Hofmann degradation of amides. Hypobromite reaction. This reaction produces an amine with one less carbon than the starting amide.

Example

$$\begin{array}{l} CH_2-\overset{\overset{O}{\|}}{C} \\ | \qquad\quad \diagdown NH \\ | \qquad\quad \diagup \\ CH_2-\underset{\underset{O}{\|}}{C} \end{array} + NaOBr \rightarrow \begin{array}{l} CH_2COOH \\ | \\ CH_2NH_2 \end{array}$$

succinimide → 3-Aminopropanoic acid

H) N-bromosuccinimide (NBS).

$$\underset{\text{succinimide}}{\begin{matrix} CH_2-C(=O) \\ | \qquad\quad \\ CH_2-C(=O) \end{matrix}\!\!>NH} + NaOBr \xrightarrow{0°C} \underset{\text{NBS}}{\begin{matrix} CH_2-C(=O) \\ | \qquad\quad \\ CH_2-C(=O) \end{matrix}\!\!>NBr}$$

NBS is a very useful reagent for the bromination of methyl or methylene groups adjacent to double or triple bonds.

Problem Solving Examples:

Amides with structures like the following are difficult to prepare and are relatively unstable. Explain.

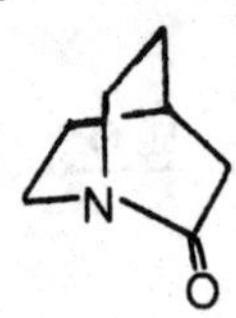

An Amide can be stabilized by electron delocation of the type:

$$R-\overset{\overset{\ddot{O}:}{||}}{C}-\ddot{N}H_2 \leftrightarrow R-\overset{\overset{:\ddot{O}:^-}{|}}{C}=\overset{+}{N}H_2$$

This type of delocalization is unfavorable in the above structure because of the high strain associated with the double bond at the bridgehead position. This high strain is due to the extremely inefficient overlap of the p-orbitals that form the π bond. This is known as Bredt's Rule and results in loss of stabilization compared to an ordinary amide. Hence, electron delocalization does not occur in the above structure.

Does Not Occur:

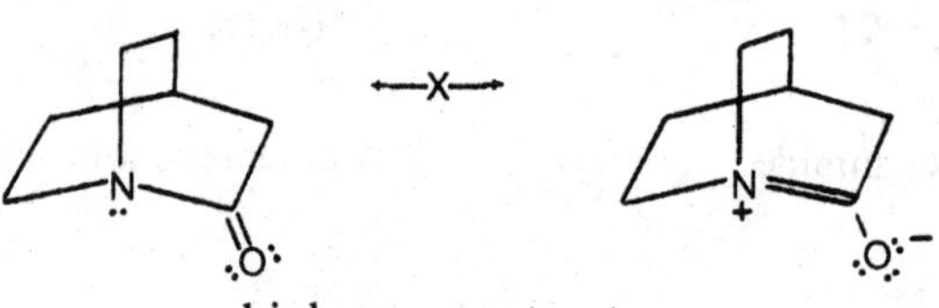

high-energy structure

Q Compound A of molecular formula $C_6H_{12}N_2O_2$ reacts with aqueous nitrous acid to give compound B of formula $C_6H_{10}O_4$. Compound B readily loses water on heating to give C, $C_6H_8O_3$. Compound A can also react with a solution of bromine and sodium hydroxide in water to give D, $C_4H_{12}N_2$, which on treatment with nitrous acid and perchloric acid gives methyl ethyl ketone. Write structures of all compounds and the equations involved.

A The key to this problem is the determination of the structure of compound A. The type of organic compound is indicated by the fact that A reacts with a solution of bromine (Br^2) and sodium hydroxide in water. A mixture of Br^2 and NaOH produces $Na^{+-}OBr$, which is the reagent used in the Hofmann degradation of amides to primary amines. Hofmann degradation can be illustrated as:

$$R{-}C(=\bar{O})NH_2 + Br_2 + 4NaOH \rightarrow RNH_2 + 2NaBr + Na_2CO_3 + 2H_2O$$

The fact that compound A, a six-carbon molecule, becomes a four-carbon molecule after the degradation indicates A must be a diamide. Consequently, D, the product of the degradation, must be a diamine. The reaction can be written as:

$$[A]\quad C_4H_8(CONH_2)_2 \xrightarrow[H_2O]{Br_2,\ NaOH} C_4H_8(NH_2)_2\quad [D]$$

$C_6H_{12}N_2O_2$

Compound A is reacted with nitrous acid to give $C_6H_{10}O_4$, compound B. This reaction is an example of the fact that unsubstituted amides (such as A) are converted to carboxylic acids by treatment with nitrous acid. Therefore, the equation of the reaction becomes:

$$\underset{[A]}{C_4H_8(CONH_2)_2} + \underset{\text{(Nitrous acid)}}{HONO} \longrightarrow \underset{[B]}{C_4H_8(COOH)_2} + 2N_2$$

B loses water to yield C, $C_6H_8O_3$. By examining B, one finds that the only way to lose water is for the carboxylic groups to form an anhydride group but B cannot be a straight chain diacid for only ß and Δ diacids undergo dehydration (2 or 3 carbons between the carboxyl groups). To write C, however, more structural detail is required in B.

This detail can be found by using the last bit of information—the fact that D yields methyl ethyl ketone after nitrous and perchloric acid ($HClO_4$) treatment. When D, the diamine, reacts with these acids, it deaminates and undergoes a pinacol type of arrangement as shown:

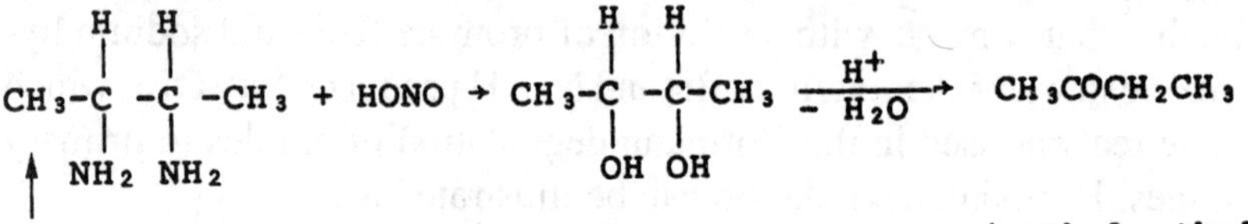

```
        H    H                       H    H
        |    |                       |    |                 H+
CH3-C  -C  -CH3  + HONO  →  CH3-C  -C-CH3  ————————→  CH3COCH2CH3
        |    |                       |    |            - H2O
  ↑   NH2  NH2                      OH   OH
  |
(Structure D)                  A pinacol)              (methyl ethyl
                                                        ketone)
```

This sequence of reactions indicates that the structural formula of compound A must be the diamide of 2,3-dimethyl-succinic acid or

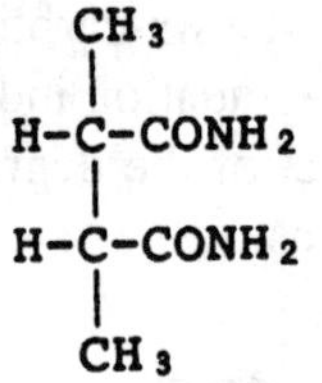

```
    CH3
    |
H-C-CONH2
    |
H-C-CONH2
    |
    CH3
```

This indicates that B can be written as

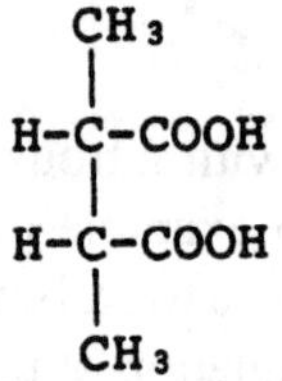

```
    CH3
    |
H-C-COOH
    |
H-C-COOH
    |
    CH3
```

which is called 2,3 dimethyl-succinic acid. With this structure, the reaction that involves the loss of water becomes:

```
   CH3                                          O
   |                                           //
H-C-COOH       - H2O               H  C
   |          ------->              |/  \
H-C-COOH       heat          CH3-C        O
   |                              |       /
   CH3                       CH3-C     /
                                  |  C
                                  H  ||
                                     O
```

C

Compound C is, therefore, 2,3-dimethylsuccinic anhydride.

Quiz: Carboxylic Acids and Carboxylic Acid Derivatives

1. $R-C\equiv N \rightarrow R-COOH$
   The conversion can be accomplished with

   (I) H_2O, H_2SO_4.

   (II) H_2O, KOH.

   (III) $KMnO_4$.

   (A) I only.
   (B) II only.
   (C) III only.
   (D) I and II only.
   (E) I, II, and III.

2. Consider the reaction below and indicate which of the following is (are) the major product(s) of this reaction.

```
          CH3
          |
CH3CH2CH CH2OH  --KMnO4-->
```

(A)
```
          CH3
          |
CH3CH2CH CHO
```

(B)

$$CH_3CH_2\overset{\overset{\displaystyle CH_3}{|}}{C}H\,COOH$$

(C)

$$CH_3CH_2\overset{\overset{\displaystyle CH_2}{\|}}{C}CH_2OH$$

(D) Both (B) and (C).

(E) Hard to tell from given information, but (A) and (B) are formed in large amounts.

3. Acid-catalyzed esterification of carboxylic acids produces which one of the following as its final product?

(A) $R\overset{\overset{\displaystyle O}{\|}}{C}OR'$ (D) $R\overset{\overset{\displaystyle O}{\|}}{C}NHR'$

(B) $R\overset{\overset{\displaystyle O}{\|}}{C}Cl$ (E) $R\overset{\overset{\displaystyle O}{\|}}{C}Br$

(C) $R\overset{\overset{\displaystyle O}{\|}}{C}NH_2$

4. Trimethylacetic acid $[(CH_3)_3CCOOH]$ is converted to ethyl trimethylacetate $[(CH_3)_3CCOOC_2H_5]$ by treating it with

(A) thionyl chloride ($SOCl_2$) followed by ethanol (C_2H_5OH).

(B) a basic solution and ethanol.

(C) $FeBr_3/Br_2$ and ethanol.

(D) dry ethanol.

(E) pure ethanol followed by heating.

5. The carboxylic acid, $CH_2=CHCOOH$, is named

(A) propanoic acid.

(B) propenoic acid.

(C) 2-butenoic acid.

(D) 2-methanoic acid.

(E) pentanoic acid.

6. Which one of the following groups is most acidic?

(A) RH

(B) ROH

(C) NH_3

(D) HOH

(E) RCOOH

7. The physical properties of a carboxylic acid change when which one of the following groups is attached?

(A) Aliphatic

(B) Aromatic

(C) Saturated

(D) Unsaturated

(E) None of the above.

8. This structure is know as a(n)

$$CH_3\underset{\underset{O}{\|}}{C}-O-\underset{\underset{O}{\|}}{C}CH_3$$

(A) carboxylic acid.

(B) acid chloride.

(C) carboxylic acid anhydride.

(D) ester.

(E) amide.

9. This reaction is an example of

$$C_6H_5CONH_2 + NaOBr + 2NaOH \longrightarrow C_6H_5NH_2 + Na_2CO_3 + NaBr + H_2O$$

(A) Friedel-Crafts acylation.

(B) Hofmann degradation.

(C) transesterification.

(D) Gabriel's synthesis.

(E) Beckmann rearrangement.

10. Reactions of carboxylic acids may yield all of the following products EXCEPT

(A) salt.

(B) acid anhydride.

(C) glycol.

(D) ester.

(E) amide.

ANSWER KEY

1. (D)
2. (B)
3. (A)
4. (A)
5. (B)
6. (E)
7. (E)
8. (C)
9. (B)
10. (C)